The Grazier's Guide to Trees is such a joy to read during these challenging times! Rather than attempting to impress us with statistics and highly technical language, Austin leads us on a conversation of discovery that makes silvopasture seem like good, ol' common sense. Thank you, Austin, for helping us to see how easy it is to create a wholesome, productive future.

—*Mark Shepard, author of Restoration Agriculture*

Drawing from experience installing silvopasture systems across hundreds of acres, Austin offers a practical, experience-based guide to planning, species selection, layout, and implementation. Clear-eyed about the challenges and grounded in reality, this book shows what's possible with thoughtful design and preparation. A valuable addition to any agroforestry shelf.

—*Steve Gabriel, author of Silvopasture and farmer at Wellspring Forest Farm*

The Grazier's Guide to Trees

How to Integrate Silvopasture for Resilience and Profit

Austin Unruh

About Acres U.S.A.

Founded in 1971 by Charles Walters, *Acres U.S.A.* magazine emerged from the need to promote ecological farming practices in a time when industrial agriculture was heavily reliant on synthetic fertilizers and pesticides. Inspired by figures like Rachel Carson and Dr. William Albrecht, Walters used the magazine, and later books and conferences, to advocate for sustainable agriculture that prioritized soil health and natural processes. Acres U.S.A. provided a platform for these ideas and helped to popularize alternative methods like cover cropping and integrated livestock management.

Though the agricultural landscape still relies heavily on conventional methods, Acres U.S.A. has been instrumental in the growing movement toward regenerative agriculture. By disseminating knowledge and supporting eco-conscious farmers, the company continues to champion sustainable practices through its publications, conferences, and online resources, contributing to a shift toward a more ecologically sound approach to farming.

Find Out More About Acres U.S.A.

Subscribe to the Online or Print Magazine
https://go.acresusa.com/magazine39896

Attend Our Eco-Ag Conference
https://go.acresusa.com/events39896

Visit the Acres U.S.A. Bookstore
https://go.acresusa.com/bookstore39896

Join the Free Newsletter
https://go.acresusa.com/newsletter39896

Photo courtesy of Gabriel Pent.

The Grazier's Guide to Trees

How to Integrate Silvopasture for Resilience and Profit

Austin Unruh

REVISED SECOND EDITION

Acres U.S.A.
Viroqua, Wisconsin

The Grazier's Guide to Trees

HOW TO INTEGRATE SILVOPASTURE FOR RESILIENCE AND PROFIT

PO Box 351
Viroqua, WI 54665 U.S.A.
512-892-4400
info@acresusa.com • www.acresusa.com

Printed in the United States of America

ISBN: 978-1-60173-989-6

This material is based upon work supported by the National Institute of Food and Agriculture, U.S. Department of Agriculture, through the Northeast Sustainable Agriculture Research and Education program under subaward number ENE23-187. Any opinions, findings, conclusions, or recommendations expressed in this publication are those of the author(s) and do not necessarily reflect the view of the U.S. Department of Agriculture.

Table of Contents

Foreword

I decided about three decades ago that I was a pretty good cattleman. I was pleased with this, because cattle were pretty much all that I was interested in.

That changed. I became interested in my farm as an ecosystem. Cattle cashflowed the operation of the ecosystem, but they were poorly positioned to be my farm's lone residents. My newly developing ecosystem needed trees and shrubs and other plants—besides the perennial grass species that I had so much love for.

I tried in vain to grow trees and shrubs, but my cattle, hogs, sheep, and goats resisted my efforts. When I met Austin, though, all of that changed. He has coached my staff and me on what it takes to have many contributory species of trees feeding our livestock.

Thank you, Austin.

—Will Harris, White Oak Pastures

Preface

I never set out to publish a book or anything resembling one.

The genesis of this guide was a series of articles I wrote back in 2020, when the potential for silvopasture plantings in actively grazed pastures was just unfolding in my mind. We had a few small trial plantings going on at the time and were trying to determine what worked, what didn't, what was practical, and how to get it funded. As I wrote, I would post articles on my website. The problem was, most of my local clients were Amish and thus didn't have online access. I printed a collection of those articles, along with some tree profiles, so they would have some basic resources to start with.

From there, I started on this guide. Initially, it was intended simply as a primer for our consulting clients, because I was getting tired of spending most of my consulting time going over the basics and was wishing I had a good resource to get folks up to speed. Then we could skip the intro and jump right into application.

And that is where it was supposed to end—a simple guide just for the silvopasture clients we took on. But the more I wrote and compiled, the more it became clear that this could be valuable to more people than I could ever hope to consult with. My driving force has always been to see more graziers unlock the power of trees for their farms, and I didn't want to be the bottleneck to that happening at scale. And by that point, I had created so much content that putting it together as something more formal no longer seemed as daunting.

The first edition of The Grazier's Guide to Trees, printed in early 2022, struck a balance between getting the information out there quickly and recognizing that, at the pace of learning, the guide would be outdated the moment it was printed. The same holds true as I write this second edition in 2025. We have learned a lot since the first edition went to print. We've refined the protocols we use for establishing trees and have steadily seen our survival rates move upwards. We have learned from the experiences of many more farmers and practitioners. And, slow and painful as it has sometimes been, we have made real progress on the nursery front, which I believe will yield the biggest advance for silvopasture practitioners of anything we've done so far. Yet there's still so much to be learned. We have gotten good at planting and protecting trees but have had precious few chances yet to study mature, yielding trees in silvopasture systems. There is a whole plethora of learning to be done once some of these systems mature, and by then what you are holding will again be out of date.

But this information is valuable as it is right now and needs to get into the hands of graziers so they can start applying it. Farmers are being squeezed by the high costs of inputs, low margins, volatile weather conditions, and a public that demands ever-greater environmental quality and animal welfare in the food they purchase. The time is ripe for silvopasture. We cannot wait for perfect, complete knowledge, because that day will never come. So I will continue planting, experimenting, and observing and will share updates as I learn. I hope you will share what you learn along the way as well. May you find this guide useful and empowering as you take your grazing to new heights.

Introduction

If you want to make small changes, change the way you do things.
If you want to make major changes, change the way you see things.
—Don Campbell

Grazing has only scratched the surface.

It has only scratched the surface because it has almost exclusively been applied to the surface—those few short feet above and below the soil line.

And yet, grazing has come a long way. The art and science of grazing have improved by leaps and bounds over the past few decades. Grazing management has been systematically improved by a host of graziers, scientists, innovators, and mavericks—from set stocking to simple rotations to, well, whatever your preferred term is for management-intensive grazing. Improvements in portable watering, fencing, shade, and coops—along with upgraded branding and marketing opportunities—have unlocked new worlds of possibilities for producers. Meanwhile, the grassfed movement has grown into a powerful force for landscape-scale soil regeneration and has become a wellspring of nutritious, wholesome food. Well-managed grass farms have transformed diets, watersheds, and communities for the better. Grass farmers have taken plowed, eroded, overgrazed, and burned-down lands and restored them back to life by providing perennial cover and reintroducing the modern versions of migrating bison, elk, and mastodons. Now, grazing is ready for the next step—up.

A next step is needed because, despite so much progress, certain problems persist and are felt by nearly every grass farmer. These problems are so persistent that they are taken as inescapable facts of life, regrettable yet as inevitable as death and taxes. Summer will bring heat stress, forage slumps, and drought. Winter brings cold stress and big hay bills. All these things can of course be compensated for, but only at real cost and effort, eating away at the financial viability of grass farms and hence their ability to sustain a joyous, simple, and rooted way of life.

The beauty is that it doesn't actually have to be this way. The solution lies not in some new chemical, shiny machine, or magic potion to spray on the pasture. The solution lies in trees. Simple, reliable, old-fashioned trees. No gimmicks, no magic bullets, no get-rich-quick schemes. Instead, for those who can muster some patience and forethought, integrating trees into a pasture operation represents a proven route to take grazing to new heights. The right trees will lower your expenses and raise your productivity, giving you more margin than before. They can diversify your income and give you greater resilience to whatever the weather and markets throw your way—all while fostering more life in the soil, birds in the air, and hope on the farm. It's about real-world dollars and cents, but also so much more. It's about making tangible improvements to the land—improvements that get better year after year. It's about investing in the long-term regeneration of a place. It's about leaving a bright legacy. What starts as a tiny seedling—no thicker than a pencil when you tuck its roots into the soil on your farm—will before long provide the shade, shelter, and feed your farm needs for decades or centuries to come.

New as it may seem, silvopasture is deeply rooted in the richness of savanna ecosystems from around the world—ecosystems that support vast numbers of wild and domestic animals. While the African savanna is the most iconic, much of the American landscape was once a lush oak-hickory-chestnut savanna teeming with bison, elk, deer, bears, and turkeys. This grazier's paradise—which was lost due to neglect, the plow, and industrial agriculture—can be reimagined and recreated through the thoughtful care of the stewards who are already committed to this meaningful work.

This abundant, profitable, resilient future is available to all who act. No patents, secrets, or restrictions stand in your way. But act you must, because trees do take time to grow and become useful. And before you act, you will need a plan of action.

The purpose of this guide is to give you the tools and insights you need to use trees—to make that plan so you can take grazing to new heights on your farm. We created it so those tools might be accessible to farmers well beyond the small radius of southeastern Pennsylvania where Trees For Graziers is located. While the species used may change the farther you are removed from our region, the principles will remain the same regardless of where you are.

Spain and Portugal have for centuries had millions of acres in a two-story agriculture that diversifies farm incomes, reduces heat stress, and adds valuable feed—all in very dry, challenging conditions. We can do the same, and better.

How to Use This Guide

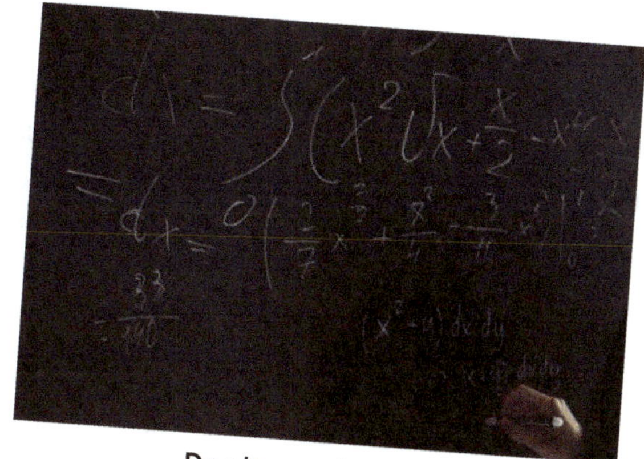

Don't overthink it!

The aim of this guide is to equip you with the insight you need to establish trees in your grazing operation. By the time you have finished this guide and the accompanying worksheet, you will have the information necessary to write a plan for how you will develop a silvopasture system. This guide—and the plan you create out of it—are for you and those you're working with. Not for me. No need to impress anybody or get all fancy. This needs to be practical and usable for you.

This ain't rocket science. It just needs to be planned and executed in a way that suits your farm.

There is no perfect silvopasture system, just as there's no perfect pasture mix or farm layout. Each has strengths and weaknesses. The goal is to determine what works on your farm and go with it.

Here's how I recommend you use this guide. Don't read it once through and expect to have a plan for your farm ready and finalized. Instead, here's what we suggest:

- **Turn first to the questions in the worksheet on page 169** and identify which questions you'll need to answer in the process of creating your silvopasture plan.
- Read through this guide once and let it digest a bit. Turn to the questions again and start to pencil in your answers. Remember: pencil, not pen.
- Return to this guide and address any blanks in your worksheet or things you're unsure of. Repeat as many times as needed.

After all that, remember that you're charting out two separate things:

- The long-term (rough) silvopasture plan for the farm
- A very short-term (detailed) silvopasture plan for your first phase of planting

By the end, you should have a good idea of what your first phase of planting will look like and what exactly you'll need to complete it. You'll want to know which species of trees, how many trees, where they're going, how to protect them, what funding you'll use, etc. You do not, however, need to know the details of what Phases 2, 3, 4, 5, etc. will look like several years from now. Those will change over time as you learn from Phase 1 and reevaluate before moving on.

You're not writing in stone. You can—and should—change and adjust as you go. This is a process, and you'll learn a lot in the process, just as you've learned a lot about grazing over the years. This is your first version of the tree plan. You reserve the right to change your mind at any time.

TWO PATHS

When adding trees to your operation, there are two main paths to take: plant trees to support livestock enterprises, or plant trees for harvestable tree crop enterprises. Both are valuable routes with their own strengths and weaknesses. However, the main thrust of this guide will focus on adding trees to meet the needs of the livestock enterprise. If you are interested in learning more about adding trees for tree crops, I suggest looking up the work of the Wisconsin-based Savanna Institute.[1]

1 The Savanna Institute's website is https://www.savannainstitute.org/.

SECTION 1
Identifying Problems and Opportunities

The first step in developing a silvopasture plan for your farm is simply to determine your goals. What specific problems need to be addressed, and which opportunities would you like to pursue? Our first focus will be on addressing issues common to all grass-based farms, because these represent the most widespread low-hanging fruit and can readily be applied at scale across the farm.

Every grass farm has real and significant problems that trees can fix. Which problems those are and how you specifically address them will depend on your context, but you can certainly solve or at least significantly reduce them. Many problems stem from either too little or too much heat. In one case, you get cold stress and a dormant season for forages, and in the other case, you get heat stress and forage growth slows to a crawl.

I like to think of trees as another set of tools in your farm toolbox that you can use to address these key issues on your farm in a way that other tools cannot. A good grazier knows how to use various grasses, forbs, and legumes to optimize their feed, but those tools can only go so far. You can't get shade from clover, and even the tallest grasses won't amount to much windbreak (unless we're talking about bamboo, which we'll cover later). Cool-season perennial forages will want to go dormant in the summer heat, and grazing 365 days is incredibly difficult without feeding hay. Yet each of those problems can be diminished with trees, once you learn how to make use of this new set of tools for your farm.

Each one of these issues can of course be addressed via artificial means, and those artificial solutions have been the status quo in grazing to date. However, artificial, manmade methods are inevitably expensive, labor-intensive, ecologically unsustainable, or all three. Shade can, of course, be provided by a portable shade shelter, but it needs to be moved frequently to avoid dead spots and it costs as much as planting a thousand trees. Winter feed can be provided by hay, but that comes at real expense, regardless of whether you make it or buy it. Barns provide great windbreaks during the winter, but the concentration of animals in the barn deprives the pastures of their nutrients unless long workdays, diesel, and tractors conspire to spread the manure back where it belongs—again at significant expense. By harnessing natural systems, you can reduce input costs and increase both your bottom line and the joy you derive from your farm.

Photo courtesy of Costa Boutsikaris.

CHAPTER 1
How Trees Benefit Livestock

HEAT STRESS

Reducing heat stress is the number one goal on most farmers' minds when adding trees to pastures—for good reason, because countless studies have shown that heat stress can dramatically reduce production and cut into farm profit. Not only does production increase with the right access to shade, but reproduction does as well, because livestock with access to shade conceive at significantly higher rates. University of Missouri researchers found that cows with shade had a pregnancy rate of 87.5 percent, while the pregnancy rate was only 50 percent for cows without shade.[2] A long-term study found that dairy cows born to heat-stressed mothers produced eight pounds less milk per day than those whose mothers had been cooled, and the next generation produced 2.8 pounds less per day because of the heat stress their grandmothers had experienced![3] These factors alone will likely cover the cost of establishing trees over time, and all the rest can be considered gravy. The low-hanging fruit in silvopasture is not actual fruit, but shade.

I wish I could tell you exactly how much value the addition of shade adds to a farm, but unfortunately there are just too many factors to consider. If you already have shade throughout much of the farm and just need trees in a few more paddocks so you can graze those on hot days, the benefit will be less than someone building a grassfed beef operation on a former corn farm in Iowa. Your choice of black, red, or white cattle will have an effect, as will their adaptability to heat. And, of course, how hot and humid your climate is plays the biggest role. What I can say is that cattle are most comfortable at temperatures well below where you and I are comfortable. So if you are hot walking out in a pasture, you can be sure your cattle are as well. And if you don't feel like eating more than watermelon and popsicles on a hot day, your cattle probably won't be all that excited about filling up their bellies and putting on weight either.

The Temperature-Humidity Index (THI) is a common way to gauge when to expect heat stress levels in cattle. Generally, you'll start to see heat stress around a THI of 72. But THI critically does not factor in sunshine—a huge gap for folks raising livestock on pasture. Research done at the University of Nebraska-Lincoln found that, to account for the factor of sunshine, the rule of thumb is that with every 10 percent decrease in cloud

2 "Shade Aids in Cattle Growth, Producer Profit," Farms.com, May 19, 2010. Available at https://www.farms.com/farmspages/enews/newsdetails/tabid/189/default.aspx?newsid=30611.

3 J. Laporta, F. C. Ferreira, V. Ouellet, B. Dado-Senn, A. K. Almeida, A. De Vries, and G. E. Dahl. "Late-gestation Heat Stress Impairs Daughter and Granddaughter Lifetime Performance." *Journal of Dairy Science* 103, no. 8 (August 2020): 7,555–68. https://doi.org/10.3168/jds.2020-18154.

cover, your THI number should increase by 1.[4] So on a nice 72-degree day with only 30 percent humidity (THI 67), there should be no stress if the weather is overcast and moderate stress (THI 77) if it's a blue sky day. If it's a day that you and I would find hot—say 85 degrees with 60 percent humidity—the full sun would push the THI all the way to 89, which is the border of severe heat stress for cattle.

Even on the cloudy 70-degree day when this picture was taken, cattle still sought out shade. It's also common for them to seek shade on warm sunny days in the early spring when they still have their winter coats.

In one study, researchers at the University of Kentucky found that cattle given artificial shade during a relatively mild period from early May to early June had significant improvements in weight gain. The stock given shade gained an additional 1.25 pounds per day for cows, 0.41 pounds for calves, and 0.89 pounds for steers.[5] For mature cows, it was less about gaining more weight and more about not losing condition. Those without shade lost about a pound a day, while those with shade gained a quarter pound per day back after calving.

These numbers are especially significant because the maximum temperature recorded over this time was only 84 degrees, and differences at warmer times in the summer would probably be higher. Average monthly temperatures in Lexington, Kentucky—the nearest city—are hotter in June, July, August, and September than they were in May when this study was conducted. It seems likely that the heat stress differences would be at least the same, if not greater, for five months out of the year in that location.

Other studies have shown that weight gain is higher under tree shade than artificial shade, likely due to the cooling effect trees have when they transpire. Research at the University of Arkansas tracked weight gain for dry Brangus cows and found the following average daily gain on bermudagrass pastures: 1.47 pounds with no shade, 1.81 pounds with artificial shade, and 2.34 pounds with tree shade.[6]

It's worth diving into the numbers here to figure out how valuable shade can be. Shade is nice, of course, but is it valuable enough for me to go to the work and expense of providing it? The effects of heat stress on livestock are complex and incredibly dependent on a whole suite of conditions. So we'll keep the numbers simple and recognize that we're taking a stab at it with relatively few data points. Your mileage will vary.

For this simplified scenario, let's assume you have thirty steers on thirty open acres of pasture in Kentucky, with no shade. Each animal will need 25 square feet of shade, or 750 square feet for the herd. Let's also assume grazing on a thirty-day rotation, where they have access to one acre each day, and you want enough shade in each of those acres for each steer to have access at any point.

4 Shane Davis and Terry L. Mader, "Adjustments for Wind Speed and Solar Radiation to the Temperature-Humidity Index," *Nebraska Beef Cattle Reports* 224 (January 2003). https://digitalcommons.unl.edu/cgi/viewcontent.cgi?params=/context/animalscinbcr/article/1223/&path_info=Davis.pdf.

5 "Shade Aids in Cattle Growth, Producer Profit," Farms.com, May 19, 2010. https://www.farms.com/farmspages/enews/newsdetails/tabid/189/default.aspx?newsid=30611.

6 "Shade Aids in Cattle Growth."

Now, how many trees will you need to provide that shade? That is, of course, a moving target, which will complicate this exercise. A mature, open-grown tree might have a sixty-foot-diameter canopy, which would shade over 2,800 square feet, but you'll have to wait twenty years to get there. To complicate things further, trees are not flat surfaces like a shade cloth would be. They are three-dimensional, and the height of the canopy makes a significant difference because, outside of the tropics, the sun is never directly overhead. We know that even the shadow from our own bodies can be dozens of feet long in the early morning or evening. So, while the amount of shade cast by a shade structure will remain relatively constant throughout the day, the shade from a tree will be much larger in the mornings and evenings and be the smallest at solar noon. The difference is very significant, as Table 1 shows.

Table 1: Shadow area comparison chart for Lexington, Kentucky on July 21, comparing a thirty-foot-diameter flat shade circle ten feet high to a spherical tree ten feet high. This date was chosen for peak heat.

Time	Sun elevation (degrees)	Area of shade for circle (ft^2)	Area of shade for sphere (ft^2)	Percent difference
10:00 a.m.	38.7	1,295	3,766	191
12:00 p.m.	61.3	965	1,296	34
2:00 p.m.	72.1	859	912	6
4:00 p.m.	56.0	1,025	1,602	56
6:00 p.m.	32.8	1,438	5,437	278

In this case, one tree with a thirty-foot canopy diameter could very comfortably provide enough shade for all the steers at once, even when the sun is at its highest. If we assume that each steer gains an additional 0.89 pounds per day because of this shade and that holds true for five months when they would otherwise be subject to higher heat stress, then that one tree contributes to an additional gain of 26.7 pounds per month, or 133.5 pounds over five months. If cattle are going for even $1.50/pound live weight, that's an additional $200.25 worth of income per tree! For all thirty trees and thirty head of cattle, we'd be talking about an annual difference of over $6,000.

Given how valuable shade is, getting it established quickly will yield real financial returns. In my experience, fast-growing willows, poplars, or black locust can all reach twelve inches in width in under five years if you start with nice stock and give proper care. Four of those trees will give you just enough shade at the highest point in the day and plenty of shade when the sun is lower. Given that those species are cheap and easy to establish, each one should cost you less than fifty dollars, including trees, shelters, and your labor.

I should point out that more trees are better, because they spread out the shade and livestock impact. Fewer trees lead to greater chances of compaction, stress to the trees, and even tree mortality. For instance, we observed a black locust with lots of compaction under it, even though there were many other trees nearby, because it was the first to offer significant shade in a pasture and also shelter from a rainstorm that rolled through. In that rainstorm, the cows huddled under the tree and pugged it up pretty good. So the more options you have, the better. But those effects are harder to quantify than the cost of each additional tree. Once you have enough shade for all your livestock in every paddock, the value of the shade from each additional tree gets smaller and smaller. Here's where trees for shade and trees for feed differ, because with trees that drop feed or offer high-quality leaf fodder, each additional tree will provide additional feed—and they help distribute shade in the process. It still doesn't make sense to plant two hundred or five hundred honey locusts, for instance, because they will very soon crowd each other out. But you can easily plant thirty or more before your returns really start to diminish.

This is where I'll occasionally have an old-timer scowl at me and point out that trees are bad for pastures since cattle just camp underneath, trampling and manuring in one spot until inevitably the tree dies. This can, of course, be true, but it is a management issue rather than an issue with the tree itself. It's like saying that cows cause muddy streambanks or that they denude pastures. Much like, "it's not the cow, it's how," the landowner in question should say, "It's not the tree, it's me." In my area, maple trees planted in continuously grazed pastures are a common sight. However, those maples branch out at just six feet high and are often fifty feet wide, with a thick, dense canopy above them. While they give nice fall colors, they also shade out the forages beneath, and because of how short and wide the trees are, the shade hardly moves. This means that the shade, and hence the animals seeking relief from the sun, is right underneath the tree all day long—leading to the problems we're all so familiar with.

The height of the canopy makes a tremendous difference in how much the shade from the tree, and hence livestock, will move throughout the day.

Instead, we want to create a scenario where trees have their canopies as high as we can reasonably get them so that the shade moves as much as possible throughout the course of the day. This makes it possible to have livestock enjoying fresh grass while in the shade of a tree that might be twenty or fifty feet away, depending on the size of the tree and time of day. The most extreme version of this is the sycamore, a fast-growing yet long-lived tree that regularly gets eighty feet or taller with a high canopy, which could probably provide all the shade an acre needs with just one tree and a ten-year wait. More trees are better to spread the shade around, but a high canopy will do a lot to move that shade for you.

Trees won't give much shade for about three years at the earliest, so go ahead and get a mobile shade structure, but also plant your trees. In several years, you won't have to move the shade around and you can sell the shade structure.

Something we often see on grazing operations is that folks create very large paddocks so that cows have access to shade somewhere. While this may provide relief from the heat, it concentrates all that manure near the tree line and puts stress on cattle, who have to walk longer distances from shade to water or forage. That extra energy expenditure is doubly taxing in extreme heat and can be avoided with well-distributed shade.

Another regular objection is that you'll lose forage if you plant trees. There is of course a kernel of truth here, since you cannot grow grass in the deep shade of the woods. But nobody here intends to create a forest. We want a balanced amount of sun and shade, with shade that moves throughout the course of the day so that it doesn't affect any one area too much. And in many cases, we're using trees with dappled canopies that allow a lot of light to reach the forages beneath so that the forages are still getting a healthy dose of sunlight, even when they're in the shade.

One often-overlooked reality is that it's not just livestock that experience heat stress—forages do as well, particularly the cool-season forages that make up the bulk of pasture in most of the eastern US. Heat and moisture stresses are the main reasons we reliably see summer slumps in forage production, despite the ample availability of sunlight during the summer months. From pasture walks I've taken in mature silvopasture systems, there seems to be a universal agreement among practitioners that forage does better in the summer with a moderate amount of shade than in open pastures. On some of our older sites, it seems that the forages that do best are those that get shade in the early morning so that the plant can make better use of the morning dew, which is often the only moisture received for weeks at a time during summer droughts.

This shows just how much light can still come through the canopy of a honey locust or other species with a sparse canopy.

Some of the best work on the topic of shade in pastures has been done by the folks at Virginia Tech, where silvopasture systems have been researched for several decades. Two studies are particularly informative for us.

The first study we'll look at was simple: Make some big frames with slats to produce shade at different levels, put them in a pasture, and measure the forage production under them. What they found likely won't surprise you. At 70 and 50 percent shade, forages didn't grow as well as in the control (no shade). However, at 30 percent shade, annual forage production was the same as for forage out in the open—no yield reduction, despite less available sunlight. Interestingly, in the 30 percent shade treatment, the researchers measured somewhat lower yields in spring and fall compared to the open plots, but higher yields during the summer. In other words, they saw more steady forage growth throughout the year, with fewer peaks and slumps in production. In a pasture setting, where spring typically offers too much forage and summer too little, moving some production from spring to summer is exactly what we want.

The second study we'll look at is even more useful, since it looks at an actual silvopasture system with trees, as opposed to an artificial system created by using shade structures. This means there were more real-world factors at play, like competition for water, tree-induced microclimate, etc.

The setup was as follows: Forages were regularly harvested from designated plots in a seven-year-old silvopasture system where trees had been planted at high, moderate, and low densities. While no precise values were given for exactly how much shade was being cast in each treatment (say 30 versus 50 percent), the "low" shade treatment had very few trees and was very similar to an open pasture. The forage grown in each of the treatments was harvested and measured over the course of two growing seasons, then compared.

These results might well surprise you. Forage production actually increased under moderate levels of shade—and not by a tiny amount. Across both years of the study, forage yields were 16 percent greater at medium density (6,130 kilograms per hectare) than at low shade density

This red maple shows a significantly denser canopy than the honey locust, which will be less desirable for forage growth, though more cooling to livestock.

(5,280 kilograms per hectare). The researchers believe that high temperatures hurt production in plots with too little shade, while the deep shade meant nice temperatures, but too little light. Right in the middle, they found a happy Goldilocks zone.

This may not seem intuitive if you were taught that plants are all just ruthlessly competing for resources. But nature is complex. Besides cooling the ground, shade also reduces evaporation from the soil, meaning water is retained for longer. Dew stays on plants longer under shade, giving some access to water even when the rains don't come. While some of a tree's roots do indeed compete with the roots of grasses for water, a good deal of their root system is deeper than grass roots. There is also the process of hydraulic lift, whereby trees can transfer water from deep and moist soil horizons to shallow soil in drought times, enabling forages to tap into the water. The trees don't do this out of altruism. If the soil around their fine roots at the soil surface was bone-dry, those roots would shrivel and die, so the tree has incentive to move water towards the surface. But whether through altruism or self-interest, the result is that more water is kept in those upper soil horizons, allowing forages to hold on longer in droughts.

Given that heat stress is almost a universal phenomenon for us, livestock, and forages, having a well-dispersed system of trees offering dappled shade will make significant impacts on the ability of your farm to thrive, no matter what temperatures are thrown your way.

COLD STRESS

Windbreaks are probably one of the most undervalued of agroforestry practices, including by me, mostly because their effect is so challenging to quantify. And yet anyone who's been out walking on a cold windy day knows the sweet relief of getting behind a solid object, be that a barn, a truck, or a tree. And many livestock don't get the luxury of sleeping inside on howling nights.

Cold, wind, and rain require that livestock burn a lot of energy to keep warm, and any reduction in the amount of energy they have to burn will show up in reduced feed costs. In cases where livestock face significant cold stress, energy will be shunted away from growth or milk production towards simple maintenance. For example, according to a 1943 report from the Alabama Polytechnic Institute, "During a 3-year experiment, a group of cows that had access to shelter lost an average of 46 pounds each in the winter, while a similar group that received the same kind and amount of feed but provided no shelter lost 104 pounds each." This shows that, even in relatively warm southern regions, cold stress can still be a significant issue. Folks further north should see even greater effects. While livestock have an impressive tolerance for cold, wet and windy conditions can have a dramatic effect on their ability to keep warm.

Different operation types will tend to have different needs for windbreaks. Dairy operations tend to have plenty of shelter options for cows during inclement weather, whereas beef operations tend to have much less infrastructure and must rely more on windbreaks established in the field. I've been told that in Ireland, where sheep are a staple of the agricultural landscape and winds and rains can be fierce, windbreaks and hedgerows are the most important part of agroforestry and are especially valuable for protecting young stock in the first, most vulnerable days of life.

Windbreaks are a common sight throughout much of the prairie states, situated around farms to reduce the burden of ever-present winds. Yet they should be much more widely used on our farms. Conifers are very effective, but are also a one-trick pony, since they only provide windbreak and no significant browse and their shade is denser than we'd ideally want. Rows of shrubs are less effective at slowing wind, but do double-duty by providing browse, which, depending on the species available, will likely offer medicinal qualities. And since windbreaks tend to be spread throughout a farm, animals can have frequent access to trees and shrubs so that they can self-medicate as needed. In regions that routinely experience a lot of cold and wind, a strong coniferous windbreak is probably the way to go, especially for operations that require stock to spend most of their time outdoors. For those with less cold exposure, deciduous shrubs that do double-purpose as browse might be a better fit. In all cases, multiple rows do better than a single row. There's even significant research showing that windbreaks can increase yields in crop operations, suggesting that pasture forage might also grow better in such conditions.

Another option is to develop windbreaks using excess materials at hand from other silvopasture work. One of the main challenges when conducting a silvopasture thinning in forest is disposing of the tops that are too small for timber. Once you've taken all the firewood you could ever want, there's still usually tons of branches left that would need chipped (expensive and time-consuming) or piled. Brett Chedzoy, a forester from New York, recommends "slash walls," whereby tops are arranged in large, piled walls around an area to exclude deer for tree regeneration. The same idea could be applied in windbreaks. While

A living barn at Brett Chedzoy's farm—a great shelter during inclement weather.

they may not be the most efficient use of space, if the piles will be there one way or another, you might as well use them to provide a top-notch windbreak for the years to come while the pile slowly decomposes.

A brush fence gives you a productive place for woody prunings, as well as a windbreak.

A similar option that doesn't require quite as much wood is to create a brush fence. In this case, you start off with posts pounded several feet from one another in a row and stack branches and brush between the posts to form a thick fence. This would go together perfectly with pollarded trees, since it provides a productive use for the branches that are left behind after cattle have stripped them of their leaves. What otherwise would simply be a chore to clean up now becomes the productive act of creating a shelter for livestock during inclement weather, which will make a real difference in feed costs and in weight gain.

SUMMER SLUMP

There are two main reasons for planting trees for summer fodder. The first is the summer slump—that time almost every summer when many expect to feed hay because their forages are heat- and moisture-stressed and no longer productive. The other main reason is drought insurance. While droughts are less frequent than the annual summer slump, they can create more damage. Think of it as catching the cold every year versus getting hospitalized with the flu. Trees have always been used as a feed of last resort in severe droughts, yet usually without any thoughtful plan. This reality is captured perfectly in Beth Hoffman's insightful book, Bet the Farm, where her father-in-

A slash wall is a way to make use of debris from logging that would otherwise simply be piled together. It can be used to exclude deer from regenerating woods, or as a windbreak

law relays the story of a severe drought that hit the Iowa farm in his childhood:

> They went without rain for so long that there was no grass at all for the cattle…Leroy recalled how the men walked the cows down to the creek about a half mile away each day for water. "And they would cut down a tall tree every day, so the cattle could eat the leaves," Leroy told us. Feeding the cattle tree leaves, he reported, saved the herd and saved the family from financial disaster.[7]

By planning and planting ahead, you can create the resilience on your farm to weather the summer slumps and droughts that are guaranteed to come sooner or later.

This picture was taken at the farm of Steve Gabriel, author of the book Silvopasture. *He largely got into the practice of feeding tree fodder because of a severe drought that forced him to get creative about feed. Once he realized he could keep his sheep on fodder growing for free on (often invasive) trees and shrubs along the edges of his farm, he was hooked.*

Let's distinguish here between tree fodder and leaf hay. The former is green leaves used fresh from the tree or shrub, whether browsed directly or cut down from a tree. Tree hay would be the same, but bundled, dried, and stored. While it is a historical practice, I can see very little use for tree hay in the modern era. It was one thing when grass or tree hay both had to be made by hand, but now that regular hay is so mechanized, I cannot see a use for the much more labor-intensive tree hay, other than maybe feeding it to certain niche animals or using it as a treat or supplement. Let's just focus on tree leaves consumed fresh.

Given the labor involved in storing it, tree fodder will likely never replace hay in winter, which is usually when the bulk of hay is used. But it is rather ideal for summer use, because trees stay green all through the growing season and are ready to be cut for fodder at a moment's notice. Having a block of fodder trees is much like having a bunch of hay bales positioned out on pasture ready for a drought, except that the trees provide shade and habitat while they are at it, won't rot, and don't need a barn or plastic wrap. Whatever is not needed during the summer season could be used in the fall, decreasing pressure on forages and allowing them to be stockpiled for winter. That way, the more tree fodder you have available during the growing season, the less hay you have to feed in the winter—no drying of tree branches required.

The most interesting work on tree fodder is happening in tropical and subtropical regions and is often called "intensive silvopasture." Many are seeing rather dramatic increases in forage production, including two to four times more livestock carrying capacity per acre.[8] I do not know whether we in temperate regions can come anywhere close to replicating what's been done in tropical climates. One key difference is that ruminants grazing cool-season perennial forages tend to have plenty of protein, with energy as the primary limiting factor, whereas in tropical conditions

---

7 Beth Hoffman, *Bet the Farm: The Dollars and Sense of Growing Foods in America* (Island Press, 2021), 220.

8 César A. Cuartas Cardona, Juan F. Naranjo R, and Ariel Marcel Tarazona Morales, et. al., "Contribution of Intensive Silvopastoral Systems to Animal Performance and to Adaptation and Mitigation of Climate Change," *Revista Colombiana de Ciencias Pecuarias* 27, no. 2 (April 2014): 76-94. https://doi.org/10.17533/udea.rccp.324881.

protein is much more of a limiting factor, which the tree fodders offer as a valuable supplement. What is clear is that trees have a significant and valuable ability to provide vigorous fodder during times of year when pastures reliably slow down from heat stress—a tool just waiting to be used.

DORMANT SEASON

Feeding livestock during the winter is the most reliably expensive part of raising livestock on pasture. Whether you make or buy hay, it's expensive, and there's a reason you're advised to "kick the hay habit." In winter we look to "mast" trees, or trees that drop feed in the form of pods, fruits, or nuts, to supplement the forages we've been able to stockpile. Some will drop early in the year—say from June through October—providing an energy boost to your livestock and removing some pressure from your forages so they can stockpile better. Others, like honey locust and persimmon, will drop late in the season after forages have ceased to grow, greatly extending how long you can graze and potentially putting 365 days of grazing within reach.

Grass farmers are fundamentally in the solar business—riding the seasonal wave of solar energy during the spring and summer, descending in the fall, and hoping they have enough to coast through the winter. This might be the most fundamental challenge in grazing. Crop farmers get around this through grain, which is essentially a nugget of stored solar energy that, when dried, can be kept for years until needed. Think of it as a battery that can be tapped when the sun isn't hitting a solar panel. Hay represents the most common way for grass farmers to lock up solar energy for winter, but it requires a good bit of labor and machinery expense to cut, rake, bale, store, and then feed. A honey locust tree is essentially taking solar energy, transforming that into stable feed energy in the form of a pod (which is slow to rot), and then dropping it in the late fall or winter when it's most likely to stay good so that livestock can harvest the feed energy on their own, without you lifting a finger or starting up a tractor. That deal is made even sweeter when the solar energy the tree has collected would have been too much anyways for the cool-season forages below. The presence of the tree makes the forages and livestock do better while spreading out the seasonal availability of feed.

Winter feed is one of the most significant and reliable costs on a farm, and it pays to reduce it considerably.

I consider honey locust trees to be the number one feed source for supplementing winter stockpiles because their pods are basically designed to last as long as you need for grazing. They have a thick outer coat that protects the inner sugars from exposure and drop when temperatures in most areas are hovering around freezing. Especially if they can fall onto a thick stockpile of forages, as opposed to muddy ground, these will stay in great condition until stock are rotated back around. Late-dropping persimmon cultivars are probably the second-most valuable, followed by a suite of other mast crops including acorns, chestnuts, apples, pears, and early-dropping persimmons. The general rule of thumb is that the earlier in the season something drops, the quicker it will go bad, and hence the more challenging it is going to be to use in a rotational grazing system that has long rest periods before returning to an area. While this can be overcome through some creative adaptive grazing, it does present a real challenge if you want to make the most use of those crops.

It is worth noting that even tree crops that drop before winter—and thus don't technically count towards the winter stockpile directly—can reduce the pressure on forages during the fall months just like tree fodder, allowing those forages to be stockpiled more effectively for winter use. As a rule, tree crops will be high in energy and relatively low in protein, which works especially well in the eastern US where energy is the major limiting factor in most cool-season forage systems.

When considering all of this, the key is to look at when your farm needs the most additional feed and plant with that in mind. It's a real strength to be surrounded by an abundance of feed at times of year when everyone else is low. It even opens up some intriguing options of buying low and selling high. Rather than playing the same game everyone else does and trying to guess when the market will be up or down, build in strength during times of year that you can virtually guarantee will see livestock sold cheaper. When drought strikes, hay prices go through the roof, and everyone around you is destocking at bargain prices, you can cash in if you have groves of trees ready to offer fresh fodder, whether you're buying animals or offering custom grazing. When everyone else is dumping animals onto the market in the winter so they can save on hay, you can pick them up and put weight on for cheap if the ground is littered with honey locust pods on a thick layer of stockpiled forages. And if your stockpile has more than enough protein for your livestock class, you could even increase your carrying capacity (and total profits) by buying cheap hay, knowing that pods will fill the gap with their abundant energy.

By understanding how trees can provide feed at critical times of the year, you can not only address the challenges of seasonal fluctuations in forage availability on your farm, but turn your weakness into a perennial, reliable strength that can undergird your farm for generations to come. If you get it now, you'll have a decade's head start compared to those who need to see someone else succeed before they follow.

CHAPTER 2
Tree Crops for Human Consumption

Now that we have gotten an overview of how silvopasture can address the most common existing problems on your farm, let's look at how trees can open up new opportunities for you as well. While you'll see that addressing the basics of heat stress, cold stress, summer fodder, and winter stockpiled feed will represent the vast majority of gains for most farmers—and certainly are the most reliable winning strategies for large-scale application on tens of millions of acres of farms—one of the truly exciting things about silvopasture is that it can open up whole new worlds of opportunity that previously didn't exist on the farm. As Joel Salatin would say, we're stacking new enterprises onto a farm and creating multiple businesses where previously there may have been just one. For those with a creative or entrepreneurial bent—and patience to match it—this is where things really get fun.

So much of this is made possible by the fact that a well-executed silvopasture can truly be additive to a farm, without subtracting anything. The livestock operation can benefit from the shade, and on top of that, silvopasture can produce feed for livestock, food for humans, food for wildlife, timber, and more. What's more, these trees don't even have to be on your land to produce for you. If you're really creative and have a trustworthy neighbor, you can partner with nearby landowners to get trees established on their land. They get free shade, and you get whatever crops come from the trees. Like so many things in silvopasture, it's a win-win.

Now, let's look at these new opportunities one by one, recognizing that we're probably just scratching the surface.

TREE CROP ENTERPRISES

Let's start with where many folks want to start: Creating a tree crop orchard. This is also where I urge folks to apply a generous amount of prudence before jumping in, as I'll explain.

Chestnuts are the go-to species when looking to build a profitable tree crop enterprise because demand currently outstrips supply, leading to solid profitability at the moment. Currently, the US imports 7.5 million pounds of chestnuts, which sounds like a lot until you do the math and realize that's equivalent to the production from roughly four thousand acres. That's not even a blip on the radar of US agriculture, where we grow about ninety million acres of corn annually. This amount will certainly all be planted in the next few years, if it hasn't been already. Several companies are jumping in and have already planted hundreds or thousands of acres of chestnuts.

Now, I'm confident that folks can find new (and revitalize old) uses for chestnuts. Demand right now is very low relative to European countries, and chestnut roasts, festivals, and better, fresher supplies should significantly increase de-

Chestnuts are a reliable producer of carbohydrates and a valuable addition to a farm. Whether the markets for them will be reliable is another question.

mand. Chestnuts can be processed and turned into a sweet, gluten-free flour for inclusion in all kinds of recipes and shelf-stable products, thus further expanding demand. But how much is that going to be? Even if we use ten times the currently imported amount of chestnuts, we're still only at 40,000 acres. I doubt that we'll ever hit per-capita demand like there is in Europe, since we don't have the traditional cultural ties to chestnuts they do.

It's exciting to see farmers make good money on a product, and the math looks great. Two thousand pounds per acre multiplied by four or even six dollars a pound makes a lot more money per acre than just about anything else you can (legally) grow on a broad acreage. It sure beats the socks off of growing corn. But the real question is, will prices stay anything close to that high once many others get in the game?

Here's the concerning sequence I can see potentially unfolding:

- Folks see an opportunity to make money. At current prices, there's good money to be made. There's gold in them there hills!

- Lots of folks decide they want in on the action and plant orchards, believing they'll be making bank in ten to fifteen years once the orchard is in full production—

- Except that a lot of people get in, and because of the significant lag time between planting and harvesting, there's a big feedback delay in the market. The price stays high for the next several years, because hardly any new production has come on board yet. In fact, prices might go higher because the profile of chestnuts is going up, and everyone wants to try this hot new delicacy. Farmers might be getting even more for their product, causing even more growers to get in on the action.

- Then those orchards start coming online. At first, prices stabilize, because there's more supply and those with supply are boosting demand by marketing their product. Everything is great for a few more years as the early movers make really good money, which will probably lead them and others to plant even more acres. But as more and more acres keep coming online, the demand can't keep up. Prices start to go down, and then they go down some more.

The question is how far prices would likely drop. These things tend to level out eventually, usually after a crash where many abandon or rip out their trees. They'll likely level out at a place where folks selling to a commodity market hover right around break-even, as in many parts of the ag industry. Unless you differentiate yourself or create great markets, you'll probably see only marginal profits in the long term.

There will be a chestnut industry, but it likely won't make very many people rich, and not for very long. My guess is that there will be two main groups who do well. One will be those who are fairly small and can sell directly to their customers—especially through local channels, U-pick, chestnut roasts, farmers markets, etc. This includes small folks who do high-quality wholesale to local grocery stores that emphasize the freshest local produce. The second group will be those that figure this out on a large scale and are well-organized and vertically integrated to manage, harvest, process, and market efficiently. Those will also likely be the ones who can purchase from other growers who are not so vertically integrated, putting them in position to set the price. Those large, vertically integrated producers will have a leg up when it comes to selling chestnuts as an ingredient, like chestnut flour to a gluten-free food brand or whole, fresh chestnuts to a large retailer like Whole Foods. If a food company has the option to source from one big chestnut producer or from ten different small farms or co-ops with varying quality standards, prices, and sales managers to deal with, I can tell you who will have the upper hand.

Commodity markets and market speculation tend to be poor games for small farms to play if they want to make money over the long term. Commodities tend to squeeze out the little guys, and speculation is a risky game. On the other hand, a U-pick chestnut orchard or fall chestnut roast might be the perfect fit.

As a cautionary tale, look at the story of California walnuts, almonds, or pistachios. Delayed gluts can be very dangerous, because they give you a long time to get in the game while the going looks good. You can put years of investment into a crop that—depending on when it starts producing—may never earn much money. There are young walnut orchards being ripped out right now, while others are only starting to come online and will likely never pay back their investment.

Chestnuts have an advantage over walnuts and almonds because they can be eaten with minimal processing and don't need processing middlemen to get to customers. That opens up the direct-to-consumer market for small producers. The downside for chestnuts is that they are a carb-rich food, whereas walnuts, almonds, and pistachios are all high in fats and protein. Fat- and protein-rich foods are generally valued and priced higher than carbohydrates, because carbs are so readily available. So although chestnuts are currently fetching a lot more than walnuts, is a chestnut really worth more than a walnut in the long run, once markets stabilize? Chestnuts are gluten free, but so is corn, and at even twelve dollars a bushel (fifty-six pounds) for organic food-grade corn, the payment to the farmer is less than twenty-two cents a pound. Chestnuts are also not shelf stable unless they are processed, meaning that either the producer or buyers will need to refrigerate or process the nuts.

So what should you do if you want to grow chestnuts or some other tree crop?

- Don't build your budgets and expectations around being able to earn six dollars per pound long-term. Especially if you sell wholesale, set your expectations much lower.

- Have a strategy for what you'll do when prices eventually come down. How can you carve out a premium, differentiated niche? If it fits in your context, how can you sell directly to consumers?

- Think about efficiency. If this is going to scale, it'll be an actual industry, where the larger players will have custom machinery for harvesting and processing. Especially if you can't develop a retail niche, think about how—or whether—you want to compete with wholesale pricing.

- What are your backup options if it simply becomes unprofitable to harvest and process chestnuts, like it has become for many who produce walnuts? Do you run cattle, pigs, or turkeys to eat the chestnuts? Sell hunting leases to clean up the deer drawn to your orchards? Even if you never sell a pound of chestnuts and pigs aren't the right fit for you, cows will gladly eat the nuts and will appreciate the shade, so the trees can still be a net gain.

J Russell Smith, author of *Tree Crops: A Permanent Agriculture*, said it well:

> Now the nuts that people eat are fine and worthy of much improvement, but a few hundred thousand acres of them would glut the market. Not so with stock food. Once we get a cow-feed tree crop established we have a guaranteed outlet, and twenty or thirty million acres will not glut the market. We would simply convert thirty or forty million acres of our hundred million acres of corn to a more profitable and soil-saving crop… Stock foods start on an honest-to-goodness basis. They don't begin five prices high like a human food novelty and then come down bumpety-bump as soon as a few carloads are produced. (Page 32).

With all that being said, integrating chestnuts and other nut crops into your operation can be very interesting—if you do it right. They offer the opportunity to add another income to the farm, a literal vertical integration, which can allow the farm to support multiple generations. Growing chestnuts as an add-on product to an existing offering seems like a great bet, as does putting on seasonal events like chestnut roasts where you're selling the experience. Those are markets where you're not subject to competition with those entities scaling up for low-cost production.

I've focused here on the topic of chestnuts because it is a hot topic in many agroforestry circles, and I want to prevent real pain and losses by highlighting the dangers of a slow glut. It'll be a different case with crops like pecans for which there is an established market and a tradition of growing pecans with cattle. The large upside of cashing in on a new crop isn't there, but neither is the downside. Two tree crops that intrigue me are heartnuts and hickories, particularly yellowbud hickories, whose nuts can be processed for oil. Both seem to have significant, untapped upsides and the ability to integrate quite well into silvopasture systems.

Heartnuts are a Japanese walnut, and certain trees develop nuts shaped like a heart, which can be clonally propagated. The flavor is very nice and mild—similar to that of an English walnut—they crack out extremely easily compared to other walnuts, and the heart shape is very attractive. Given that the part you eat is encased in both the shell and the hull, they should be exempt from food safety guidelines that require you to keep livestock out for long periods of time before harvest, as you would need to do to sell food-grade chestnuts. They also store amazingly well, actually peaking in flavor after several years of storage. That attribute makes them perhaps the best homestead resilience food around.

The main drawback I see is that their growth habit is not conducive to silvopasture because they like to grow very wide and low, so their shade will be right underneath them all day long. They are slow to get real vertical growth, so integrating livestock among them will be more challenging and take longer. The best approach is likely to force the tree to grow tall by placing it in a tube like any other silvopasture tree, even though it naturally wants to grow out rather than up. Once it emerges from the tube, it can resume its wide-growing ways. Either way, the growth form is not a deal-breaker; it's just something that needs to be weighed in the decision-making process. For those with existing access to customers, this seems like a relatively simple addition to your offering and one that needs no infrastructure or new tools to process, at least on a small scale.

Yellowbud (aka bitternut) hickories are intriguing because they offer the possibility of establishing a domestic olive oil alternative and the trees are adaptable to a wide array of soils across most of the eastern US. Their high nutmeat content, 75 percent oil content, and thin shell make this a very promising crop. While the common name "bitternut" might scare folks off, the bitterness comes from tannins that are completely absent from the pressed oil. This is a tree that is only starting to be brought out of the woods and into cultivation, and the jury is still very much out on whether it will eventually be able to economically compete with olives as an oil crop, especially given that olives have a multi-millennia head start. Yet the downsides seem low to me, since they can readily be established and maintained in pastures, and if the trees do not turn out to be profitable to manage in a decade or two, they still provide nice shade, quality wood, and great wildlife habitat—in addition to all the oil you could ever want.

Samuel Thayer writes in Incredible Wild Edibles that "although the literature often reports that the nuts of the species are not eaten by wildlife, this is false. . . .I have watched white-tailed deer and black bears eat them avidly, and they are relished by red, gray, fox and flying squirrels as well as chipmunks and deer mice."[9] This bodes well for the backup use of the yellowbud as feed for livestock—either as a primary use if the crop is unprofitable to harvest, or just cleaning up after most nuts have been picked up. I would love to see folks go big into yellowbud plantings to see whether, with the right scale and mechanization, we can create a true domestic oil tree crop industry.

Overall, adding tree crops to a grazing operation offers a whole suite of new opportunities that can seriously diversify the farm business if done in a way that fits your farm strategically. The key difference-maker here is that these nut crops in particular can be integrated with livestock relatively easily, allowing the enterprise to stack on an existing farm, whether it's your own or someone else's. And because they offer you a crop that you can harvest and then sell—as opposed to a feed crop like honey locust or mulberries that is directly consumed by livestock—they open up business opportunities that aren't there for feed trees, like working with a nearby farm to establish trees for them, harvest the crop, and develop a revenue sharing agreement. With some sweat equity, this could allow someone to get into farming at scale without debt or even land ownership—a very interesting opportunity indeed.

My main recommendation is to use prudence and long-term vision when thinking about tree crops. Do your research and talk to folks who know the field well. These are very long-term investments, not nearly as guaranteed as trees that are there to support your livestock directly. For those who create a sound strategy and hedge their bets, tree crops are an interesting option indeed.

HOMESTEADING CROPS

Instead of speculating on the future market for tree crops, we encourage all of our clients to look at tree crops first and foremost as a means of creating a world of abundance for their families and communities. Tree crops offer the opportunity to surround yourself with some of the most nutrient-dense, reliable foods in the world and have them yield year after year, with very little labor input once they are set up.

[9] Samuel Thayer, *Incredible Wild Edibles: 36 Plants That Can Change Your Life* (Forager's Harvest Press, 2017), 161.

Grass-based farms are by nature set up to produce an abundance of quality meats and other animal proteins, but don't necessarily have nutrient-dense plant foods to fill that out, like fruits, berries, starches, or oil crops.

I for one have been publicly skeptical of betting on many tree crop enterprises, yet around my own property I have dutifully planted a whole array of tree crops for home use. The key difference between growing these crops around home and growing them for sale is that I don't have to worry about how I am going to harvest anything at scale, process it for the public, or sell it. I can pick small quantities by hand, I don't need to follow commercial guidelines for processing because I am eating it myself and know exactly how it was collected, and the final consumer is me, my kids, and our friends, while any extras go to the animals. It's as simple as that. It's like the difference between growing a home garden for your own use and starting a produce farm growing a bunch of stuff that few folks eat, like bok choy or kohlrabi.

The crops that I have personally planted include staples like chestnuts, heartnuts, buartnuts (heartnut crossed with butternut), European and hybrid hazelnuts, pecans, and walnuts. All of those can do great in a silvopasture. My "berry patch" is a dense mix of mulberries, elderberries, blackberries, red and black raspberries, gooseberries, currants, aronias, serviceberries, strawberries, goumis, gojis, autumn olives, rose hips, persimmons, pawpaws, apples, poplars (for shade), and a half-dozen minor support plants.

While this wouldn't be a great place to run cattle, geese and ducks would absolutely love it, and many of the species listed could easily be planted in a silvopasture. While many of the trees are only starting to yield four or five years after planting, I have set this up in such a way that my great-grandchildren will still be harvesting from what I have planted. I don't necessarily use everything that I plant every year. There's plenty for me, the squirrels, and the cattle. But it's a great feeling to know that I always have an abundance of crops to fall back on, should the need arise. And in the meantime, these trees and shrubs are undoubtedly adding to our health by providing both highly nutritious food and a beautiful, relaxing place to spend time with family and friends outdoors and in the shade.

For anyone looking to ensure long-term and reliable access to some of the most nutrient-dense foods in the world, planting tree crops for your family use is a perfect place to start.

The colorful harvest from the berry patch in early June.

CASE STUDY
Country Sunrise Creamery

Nelson Martin of Country Sunrise Creamery was blessed when his father made provisions for him to live with his family on the farm where he was born. Recognizing that he had an opportunity that was out of reach for most, Nelson has wanted his children to have the privilege he had. But rather than acquiring new farms for them, he is creating new farming operations on the existing land base.

With this in mind, he diversified the operation by opening an on-farm market selling their own milk, cheese, and yogurt, as well as other local products. They have planted a fruit orchard and have integrated chestnuts quite extensively into their silvopasture, which they can readily sell through their existing store and the local retailers that sell their dairy products. As the kids grow older, I am sure there will be a whole wave of new farm-based businesses blossoming where previously only one existed.

Unlike most growers, Nelson direct-seeded the chestnuts in his silvopasture, and they did well after one round of replanting. That gave him the opportunity to use the genetics he wanted most for his farm by buying nuts directly from the source; he also didn't need to grow them out for a year in a nursery. It's a good option for large-seeded trees like chestnuts, oaks, or hickories, as long as the seeds are stored well and you use some kind of rodent deterrent like castor oil.

Nelson Martin of Country Sunrise Creamery believes that stacking new enterprises on the existing land base will allow more of his family to stay—and thrive—in farming.

CHAPTER 3
Hog and Poultry Silvopasture

Now that we've covered the addition of tree crops for human consumption—whether for home use or for sale—let's move into what I believe has the potential to completely upend, for the better, the way we raise pigs and poultry in this country—and bring significantly more revenue to farms in the process.

While the last several decades have seen a tremendous renaissance of grassfed beef and dairy production as farmers and consumers alike have learned of the benefits to the land, the animals, and to our own health, the same unfortunately cannot be said about pigs and poultry. The overwhelming majority of those animals still complete their lifecycles without ever seeing the sun or smelling fresh air.

I believe a significant part of that inability to bring more chickens, turkeys, ducks, and hogs outside is that we have overlooked where those animals naturally thrive and have not actively created the environments that allow them to do so. Cattle are grazing animals and thrive on open pasture. Most beef cattle spend the first half of their lives on pasture—not because that's how folks get an animal welfare premium, but because it's just more cost-effective. It's how cattle were made to thrive.

Pigs and poultry, on the other hand, are monogastrics, meaning they only have one stomach. They don't have the rumen that gives cattle, sheep, and goats the superpower to break down complex, tough, fibrous material into what their bodies need. Instead, they need more easily digested foods. In the wild, this includes seeds, fruits, nuts, tubers, insects, grubs, small animals, and leafy forages. On farms, that wild diet has been replaced by grain, which is a major input cost and supports an unsustainable annual crop industry.

What's more, while cattle are perfectly at home in an open field, modern chickens trace their lineage to jungle fowl, not prairie chickens. Go hunting for wild turkeys or wild boar, and you'll likely be spending most of your time either in the forest or at the edge of the woods.

This is where a very significant opportunity arises. I propose that we need a better model of raising pigs and poultry on pasture—one that uses tree crops to offset a large portion of grain feed in a seasonal, high–animal welfare environment that will produce meat with a nutrient density that will almost certainly blow all others off the charts.

On the tree side, we need to roll out top-notch trees with the genetics and care needed to produce ample feed throughout the growing season—starting with mulberries in the early season with their high-protein fruits (we have tested mulberry fruits with 18 percent protein) and copious yields, then progressing towards a combination of persimmons, apples, pears, chestnuts, acorns, and more in the fall. We want high-yielding female trees that all produce crops, and we want to design these systems so that we'll have blocks of tree crops yielding in sync with one another throughout the year. Picture vast orchards, but instead of all apples or citrus destined for the grocery stores, it's a smorgasbord of feed that will turn into bacon and eggs and drumsticks.

On the livestock side, this will not be accomplished with the same animals that are chosen for industrial production. Those are selected because they can grow fast, efficiently, and in a cramped, caustic environment. The limitation of raising stock on pasture and gathering their feed under trees as they drop is that those livestock need the mobility to get there. Cornish Cross broilers that get tired after three steps simply will not do in this system. Pigs will be fairly easy to swap out, as there are numerous heritage breeds that have captured an audience and are already being raised on pasture and in the woods on a small scale.

These pigs have been a handy tool in creating this thinned silvopasture and now benefit from the shade the trees cast. But how much better would it be if they stood under mulberries, persimmons, chestnuts, and oaks?

The biggest change, however, will need to be in poultry breed selection. Highly mobile chickens will lack the large breast that folks have become accustomed to. Undoubtedly, crowds will jump up and object that you just can't sell a scrawny chicken without (unnaturally) large breast meat. But we've heard that before in the beef world—that folks wouldn't eat grassfed beef because it's too lean, too gamey, and doesn't have enough marbling—and now grassfed beef is considered a premium product. What it comes down to, then, is creating a product that is significantly superior and then educating consumers about why it's so much better. In the case of poultry raised on a silvopasture diet of fruits and nuts in an idyllic setting, the marketing will take care of itself. And as far as the quality of the meat is concerned, it's helpful to remember that preferences are almost all learned, not inherent. In many parts of the world, folks prefer lean chicken to our monstrously plump broilers because of the depth of flavor and firmer texture, as well as the higher levels of collagen that make delicious, rich broths. For someone who has already turned to grassfed beef, bone broth, and cooking with tallow, this will be an easy switch.

For the production system, my guess is that in many cases folks will use a (portable) central building, which offers protection, feed, and water, with multiple paddocks radiating out from there. There's a limit to how far we should expect animals to travel from their central housing, so rather than a fifty-acre poultry farm having just one central house, it could have a dozen smaller houses. Most systems will likely be built for seasonal production, maximizing the benefit of tree mast while minimizing the need for care during the coldest, wettest, and muddiest parts of the year. Production of turkeys and geese—animals that are already raised seasonally—fits perfectly into such a model. You could finish multiple rounds of broilers in the growing season, and layers could get their start on wide-open pasture before they start laying, at which time they could be moved to a more controlled setting, either mobile coops or a central barn. Pigs were traditionally raised in woodlands (think about Appalachian pig drives), and while pickings will be slim during the winter for those that choose to farrow then, the growing season offers a whole array of food for finishing.

Such systems should fit nicely with existing ruminant operations for grassfed beef or dairy.

The trees will offer shade and fodder for the cattle and a means of diversifying through seasonal production of pigs and poultry. My hope is that this method of stacking enterprises will create economic incentives for more people to raise livestock of all types on diverse, well-managed pastures. These farms will be economically resilient, ecologically rich, and a pleasure to live and work on.

What's more, this is likely to provide a very strong additional income for the farm by offsetting feed costs. Unlike tree crops like chestnuts, heartnuts, and hickories, which are all based on niche or emerging crops, the output of these systems would be premium bacon, eggs, or chicken, all of which have a very high and steady demand. Even if your cost of production is the same, the premium you can get for this differentiated product with top-notch visual storytelling will more than likely be worth the additional investment.

> Family businesses that go into anything where scale is an important competitive factor almost always lose out to public corporations and their access to cheap capital. [Author William T.] O'Hara found one thing that could provide this wide financial margin year-in and year-out was a product whose quality was so exceptional that consumers willingly paid an above-average price for it.[10]

To get a sense of how much savings one might anticipate, the best resource I've found is the book Tree Crops by J. Russell Smith. In that book, he shares his learning from traveling throughout the country and the globe to study how various cultures use tree crops. He found that "all through the Southern States, mulberries are commonly used as feed for pigs and poultry. In North and South Carolina and Georgia nearly every pig lot is planted with these trees, and the mulberries form a very important addition to the pig's diet."[11]

Smith goes on to share a whole suite of insights from farmers and researchers throughout the region. The challenge is that all this information is based on anecdotal evidence, with no accounts of anyone having done the precise bean counting of feed and weight gain. The general rule of thumb was that "one everbearing mulberry tree is enough to support one spring pig during the fruiting season of two months or more."12 More precisely, Smith shared several anecdotes from farmers who are very specific about what they have noted, including the following two reports:

> Mr. R. H. Ricks, a farmer specializing in cottonseed at Rocky Mount, North Carolina, gives the following testimony: "I planted two hundred mulberry trees of the everbearing variety thirty-three years ago on what I regarded as waste land. They commenced having some fruit at once, but they did not have profitable crops until the fifth year. I have since planted another orchard of fifty trees. I carry fifty to sixty hogs on the fruit ten to thirteen weeks every year for the last twenty-seven years without other food, and in the main bearing season the two hundred and fifty trees would carry twice the number of hogs. Nearly every farmer has a small orchard in this section, eastern Carolina. I regard the mulberry for hog food with much favor." . . .
>
> Mr. James C. Moore, farmer of Auburn, Alabama, writes, "I never weighed my pigs at the beginning and close of the mulberry season, but think I can safely say that a pig weighing 100 pounds at the start would

[10] Allan Nation, *Land, Livestock and Life: A Grazier's Guide to Finance* (Green Park Press, 2007).

[11] Smith, *Tree Crops*, 85.

[12] Smith, *Tree Crops*, 86.

> weigh 200 pounds at the close…Three-fourths to the mulberries is safe calculation of the gain. I have had the patch about 18 years bearing. I planted my trees just 32 feet apart, and now the branches are meeting, and I have about 40 trees. I have carried 30 head of hogs through from May 1 to August 1, with no food but the gleanings of the barn and what slops came from the kitchen of a small family."

I've done the math on these systems several times, and however you slice it, they look solidly profitable. What it comes down to is cutting out a lot of your high-cost inputs to the farm and replacing them with trees that yield heavily year after year, with livestock doing the bulk of the work. And this doesn't only work in the South. The further north you go, the less need there is for shade for cattle, but leaning hard into a seasonal production model really makes sense, with trees converting the ample sunlight from long summer days into the mulberries, chestnuts, apples, and acorns needed to put weight on livestock, which can then be butchered and kept safely away in the freezer over winter while the farm operation turns its attention elsewhere.

This works particularly well with species that can procreate quickly and finish within the timeframe of a growing season. While you need to feed one cow a lot to produce just one calf, the math is very different when talking about a sow that can produce ten pigs, or a hen that can produce dozens of broilers in a year. While New England farmers cannot buy grain as cheaply as someone in Iowa, they can grow many tree crops that produce abundant feed calories. If these tree crops are paired with lush forages that experience little summer slump in the cool Northeast and seasonal dairy that produces whey as a byproduct, you have a very complete nutritional package for monogastrics being supported mostly with local,

There's no reason we in the US cannot have extensive savanna ecosystems on an even larger scale than the Spanish dehesa.

seasonal feeds—the unique result of which can likely fetch a premium price.

Currently, the dehesa region of Spain and Portugal covers roughly ten million acres of picturesque rolling Iberian hills and produces some of the most prized meats in the world, particularly acorn-finished Iberian ham. And that's done in a Mediterranean climate with very sparse summer rainfall. Almost all of the US east of the Rockies should produce better than the dehesa because it has more generous rainfall patterns, especially in the summer when trees are under the most stress. We therefore have a much wider selection of trees that will thrive here and are not dependent on oaks and their masting cycle. There is no reason why we cannot have savanna-like farm ecosystems covering ten times the area of the dehesa, with several times more productivity per acre. The main difference is that they have a head start.

It's time to do for pigs and poultry what has already been done for ruminants—put them back in ecosystems where they thrive with minimal human input. Those are not open pastures, but rather systems with well-placed trees and shrubs offering shade, cover, and abundant seasonal feed. Whether you decide to implement this for a backyard poultry flock or a hundred-acre pastured hog operation, the foundations are time-tested, sound, and, if taken seriously, have the ability to transform the lives of millions of animals, entire landscapes, and the very basis of farm viability.

CHAPTER 4
Other Opportunities for Silvopasture

TIMBER

One of the ways that Trees For Graziers really sticks out in the field of silvopasture, especially how it's mainly been practiced for the past generation or two, is that we focus very little on timber—and I believe that has been absolutely key to our success so far. I'll go a step further and say that foresters who have looked at open pastures, drooled, and tried to convince a farmer to plant trees for timber have actually held back the practice of silvopasture by failing to understand what is driving most graziers to want trees.

While growing timber in a pasture is what many people think of when they hear "silvopasture," I think it's rarely the best use of the land.

I'm not going to say timber isn't the right fit for any silvopasture. Timber silvopasture happens to be the most widely adopted silvopasture practice in the country, with large sections of the Southeast integrating pines and cattle. But TFG does not yet operate in pine country. Our genesis is in Lancaster County, Pennsylvania, where land goes for $30,000–$50,000 per acre and dairies dominate the landscape. Our clients are graziers, first and foremost. They rely on forages for their animals and have very little interest in becoming foresters or waiting twenty years for cash flow from a timber crop. They also cannot afford to take land out of production for five or more years until the trees mature enough to reintegrate livestock, given the small land bases they farm and the price of land. So, for most of our clients, quick shade is the first thing we aim for, followed by trees that will boost feed production at critical times of the year, thereby reducing feed costs and improving the bottom line.

But I believe this strategy works outside of the unique economics of our area. I was in Virginia talking with a landowner who was looking to create a silvopasture from an open pasture and had been advised to add pine for timber, but at a lower stocking rate than a typical timber planting to accommodate forages. So far, so good. But when I asked about the size of the plot, and how large an area was needed to do a profitable commercial timber harvest, the economics started to fall apart. I can't recall exactly what the numbers were, but let's say the plot was about twenty acres, and in that area they would have needed fifteen acres of solid pine plantation to make it worthwhile for a timber crew to bring their rigs in and conduct a profitable timber harvest. Given that they were planting at a much lower density, they would have needed several times that amount of land to make a timber harvest economically feasible—more land than the landowner had available to put into silvopasture. From the beginning, it's a setup for a future situation where the landowner will be stuck with trees that are marginally profitable, if at all, and will almost certainly not repay anything for the materials, labor, and time invested in them. What might have made sense on two hundred acres just doesn't work for smaller properties.

What's more, some of the biggest and easiest gains with silvopasture, especially in the Southeast, come from simply providing shade trees to reduce heat stress. With a timber silvopasture that's based on an all-in, all-out harvesting model (as opposed to high-value timber trees that can be selectively harvested while maintaining plenty of shade), you have a period of no shade while the trees get to size, then a time when there is nice shade for a decade or so. Then all the shade gets harvested, and it'll be years before you again have the benefit of shade. I would much rather see a combination of short-term shade trees planted for quick shade, mixed in with longer-term trees that will drop persimmons and honey locust pods and low-tannin acorns for generations to come. The beauty of such a system, which is focused on serving livestock directly, is that it can work just as well on one acre as on a thousand acres.

So if you're a small landowner looking to create a silvopasture, please look really closely at the economics and scales of efficiency before you move forward with a timber planting, especially a low-margin commodity crop like pines. And even if you have a thousand acres, you'll still want to look very carefully at the numbers before you move forward. I'm no forester and certainly don't have direct experience with pine plantations, but I haven't heard a whole lot of great things about their economics. Many folks throughout the country have planted pine timber, only to learn decades later that their stand had become worthless due to increased supplies of timber (because everyone else planted these trees as well) or the loss of local sawmills.

If someone wants to do timber, I highly recommend the book Heartwood by Rowan Reid.[13] It's an Australian book, so the context is very different from mine, but the philosophy of integrating trees into farms primarily for shade and shelter, with the added benefit of high-value, well-managed timber, is spot-on. Reid calls himself a "forester among farmers" but does a wonderful job translating the concepts of forestry into the farming context, both for timber used on the farm and well-managed, high-value logs that would be worth harvesting in small batches, processing on-farm (great work during the slow season), and selling to premium markets. This neatly parallels the strategies small farmers and small timber growers need to take to be economically successful. If you're a small farmer, good luck trying to make a profit from commodity corn when you're playing the same game as operations with five thousand acres. Similarly, producing commodity pine wood on your farm is likely not a winning strategy when there are companies like Weyerhaeuser that own over ten million acres of timber.

If I were to plant trees for timber, it makes the most sense to grow posts that I can use on the farm. Black locust is probably the best option here, though I can also see interesting use cases for Osage orange (longer-lasting even than black locust), mulberries, and catalpa, which are known to have solidly rot-resistant wood. These can easily be cut and processed on the farm without expensive equipment and used for something any grazing farm needs anyway, or sold to folks needing untreated posts.

---

13 Rowan Reid, *Heartwood: The Art and Science of Growing Trees for Conservation and Profit* (Melbourne Books, 2017).

Another way of approaching a timber planting is to think of initially establishing the "mother" trees via silvopasture—trees which will later seed many more. It's relatively costly and challenging to get trees established in actively grazed pastures, and it's hard to imagine many trees paying for themselves in timber when you factor in the time value of your investment. It's also harder to get trees grown in the open to have the nice, knot-free, and straight form they need for quality timber. It can be done, but it takes consistent and thoughtful care over decades. That said, if you can establish, say, twenty or thirty trees to the acre of a timber species like white oaks, black walnut, or improved black locust, they give you the option that, once established, you could decide to let the area go and those trees will seed a full, high-density forest for you. At least for oaks and walnuts it's through seed. With black locusts I would actually cut the original trees down to induce each tree to throw up dozens or hundreds of new root suckers. If you let that area go, you'll have a true timber stand very quickly.

Thankfully, even though pine silvopasture has been the most widespread form of silvopasture to date, we've got much better options available that are less reliant on commodity markets outside of our control.

HUNTING

While the perennial cool-season forages that are the basis for much of the grazing acreage in this country thrive in the spring and warm-season forages thrive in the summer, most trees spend their whole year working towards the fall. Honey locusts, persimmons, apples, pears, oaks, and chestnuts—all of which are loved by deer, bear, and turkeys—spend months converting solar energy into feed energy in the form of fruits, nuts, and pods. And what coincides with the drop of all those delicious, high-calorie foods? Hunting season. While much of any autumnal agricultural landscape might be a mixture of bare crop fields, grazed-off pastures, and overgrown woods, a well-managed savanna-like silvopasture system will rain food from the sky—while providing cover for wildlife and trees to hang a deer stand in.

The trees that drop food for cattle or sheep will work for deer as well, and in fact deer will open up uses for a much greater variety of trees in a silvopasture system. While honey locust pods drop late, are naturally quite rot-resistant, and will thus stay in good condition until livestock move through and pick them up, many tree crops that drop earlier in the season are much more perishable. Early-dropping persimmons, apples, chestnuts, pears, plums, mulberries, and others are more challenging for rotationally grazed ruminant systems because much of the mast tends to

While crabapples might not be the most productive tree for feeding cattle, deer will gladly come to feed on them throughout a long season.

rot before the livestock are rotated back around. That's not a problem at all for deer, since they will gladly help themselves to whatever fruit is falling on your property whenever it's falling.

This is the reason why silvopasture offers certain advantages to continuous grazing operations that are not available to those practicing intensive rotational grazing where livestock are excluded from certain areas for long periods of time. I know that set-stock grazing is heresy in certain crowds, but the fact that cattle have constant access to any tree they'd want and can pick up the high-energy fruits right as they drop makes for easier maximization of the tree crops. For the many producers who aren't able to do daily moves, this opens up a suite of valuable and easily accessible tree species—like apples, pears, and chestnuts—that are less useful for graziers who use long rest periods. In fact, there will be instances where it is logically easier to increase livestock yields through the addition of tree crops than through increasing the frequency of livestock rotation.

Going back to hunting, if you can help a deer-crazy hunting landowner develop his land through silvopasture in such a way that allows him to spend countless hours doing what he loves most (and maybe even bringing home a trophy buck), you'll get access to all kinds of land. Not only that, but the land will be that much better for your livestock as well, offering distributed shade and feed where there was none before. It's much like the approach Greg Judy has taken, treating the landowners he leases from with great care by installing great fences and even developing well-made ponds. This is one more way you can add value to the relationship and ensure long-term land access without the financial burden of land ownership.

Alternatively, if you own the land and improve the hunting experience, you can choose to either keep that for yourself or share that experience with others for a fee through hunting leases. While a grass farmer trying to make money might pinch every penny, hunters will drop huge sums of money on gear, guns, and, if all goes well, taxidermy—which makes them great customers. There are now several online services available matching landowners with hunters, and you're sure to stand out if you can offer a rich stand of tree crops drawing in wildlife from all around. Just make sure to keep the cows away when hunters are in.

Now, trees are indeed an investment in time and money. I wouldn't recommend that you go and plant a bunch of trees on your dime for the landowner if you're leasing a piece of property. Unlike fence posts, which can be pulled up if the lease is terminated, you can't just move those trees. So unless you have a really secure long-term contract, the burden for the planting should fall on the landowner. They could pay out of pocket or seek out silvopasture funding options like those available through their state NRCS office. It's even better if you can get paid as a contractor to plant those trees for the landowner.

As with tree crops harvested for humans, hogs or hens, the beauty of this system is that you gain the upside of the shade, shelter, and a bonus of feed for livestock, plus additional benefits that you can gear towards wildlife. It's a win-win for all involved. It can unlock great land access, new revenue streams, or just simple good memories of sitting in a deer stand watching animals come into the rich feed that you put there, just for them.

BEES AND WILDLIFE

I'm no beekeeper, but I have learned that beekeepers get really into their little winged livestock. So whenever we do a planting, we ask folks if they want to integrate trees for bees, which typically include black locust and basswood but could include all kinds of others like catalpa, sourwood, and black gum.

This might be a stretch, but it's worth noting that if you have the tree-planting bug as bad as I do and like keeping bees, they offer a unique opportunity because you can help your neighbors by establishing trees to shade their livestock and directly benefit from those trees when your little honey-makers fly over there and help themselves to the flowers. As long as the plantings are within a mile or two, you have a means of extending the reach of your "farm" without having to tie up any of your land, or even having to own land at all. If you can get paid to do the planting—say through a conservation grant—then we're really talking.

Beyond honeybees, we want all farms to buzz with life. We want wild and abundant ecosystems overflowing with vitality, with birds, bees, and butterflies all frolicking through the air while worms burrow, turkeys gobble, and fawns take their first wobbly steps. We want bats nesting in trees and swooping to snatch insects from the air, while wood ducks build their nests beside wooded farm ponds. Some of this will make money, but mostly it makes for a good life for you, me, and everyone else. You don't get into grazing because it's going to make you a ton of money. You choose to graze because you want to create an abundance of life and bask in what you've made possible. Whether you're creating nesting sites for songbirds where previously there were none or seeing turkeys thrive because they now have consistent feed year-round, this is nourishing to the soul in a way that can't be valued in dollars and cents.

Creating a habitat for life on the farm is perhaps one of the greatest legacies you can leave.

ENGAGE WITH PEOPLE!

So far, we've been focused mostly on how you can create more products from a farm—chestnuts, pork, eggs, fence posts, or even honey. But I believe that one of the most significant advantages small farms have is that they can provide unmatched experiences for folks who really want to get reconnected to the land, farms, and food. And many folks are willing to pay more for experiences than they are for food.

Any farm can engage with the public through a petting zoo, corn maze, or any number of other activities. But there's something about a savanna-type landscape that speaks pretty deeply to the human psyche and draws people in. It's beautiful, lush, abundant, and rare. Just by having the trees, you will stand out from the crowd by offering a space that is the epitome of cool. You can open up interest by hosting photo shoots, parties, or even weddings. And then the trees can take it up a level by bringing the goods needed for all kinds of events—U-pick mulberries, chestnut roasts, pawpaw festivals, persimmon pudding contests (for all of you in Indiana), cider tastings, and so much more. A lot of folks who are now several generations removed from the farm feel disconnected and want to know where food comes from—especially nutritious, whole food. If you can draw them to your farm through seasonal events, you can then sell them on your grass-finished beef, mulberry-finished chicken, or even the grassfed butter from your neighbor down the road.

When I was visiting southern Florida, our guide told us to visit a famous fruit stand called "Robert Is Here." While there are thousands of farms growing tropical fruits in that part of the

This is the kind of farm that draws people in to see what's going on.

state, nobody offers an experience quite like Robert Is Here. Ever since 1959, when a young boy named Robert first manned a roadside fruit stand, it has been built out to a real tourist attraction, complete with more types of tropical fruit than you've ever imagined or heard of. It has milkshakes, tons of preserves, and even a petting zoo. And the staff really leans into making it an experience, like when I got a five-minute consultation on how exactly to know when a black sapote is ready. I'm sure I could have gotten all of these fruits cheaper elsewhere, but to me, the experience was well worth it.

While southern Florida is a unique place with a lot of tourists moving through, I believe that most farms in an area with enough traffic could replicate something similar, highlighting unique and seasonal crops from the area either full-time or just in the form of seasonal events that draw people in. In fact, really leaning in and providing quality experiences rather than just farm products is probably most effective as a farm strategy in areas with the least ability to compete on regular farm production, because they are often close to metropolitan areas with sky-high land prices and small parcel sizes. What seems like a weakness when farming "the way everyone else does it" becomes a real strength when you look through a different lens.

And you don't have to go to Florida to see examples. I can get to several orchards within a half-hour drive, even though my area is not a major commercial fruit-producing center like Washington, Michigan, or even central Pennsylvania. What we do have are a lot of people who live nearby or drive through. These places draw people in with petting zoos, hayrides, farm festivals, and U-pick and attract a crowd willing to pay a premium for the experience of it all.

I recognize that this is not for everyone and that many people get into farming precisely because they don't want to deal with people. But for those extroverts out there, creating an engaging experience for the public is likely one of the best ways to make money from the farm. This might be just the opportunity for someone who is currently off-farm but would love to bring their talents in marketing or photography or wedding planning back to the farm, while others keep focused on producing the same goods they always have, now for a higher retail price.

TREE PLANTING BUSINESS

I have a particular passion for seeing more people get into the business of tree planting, because it is absolutely key for silvopasture to thrive and spread throughout the country. My goal is that any farmer in the United States will have someone in their area that they can go to when they want to add trees to their farm. We have been able to spur on silvopasture plantings so much that we now have, by far, the highest density of planted hardwood silvopastures in the country in our little region of southeastern Pennsylvania. I am confident that, had it not been for someone like us making this simple for folks, there'd be less than 10 percent of what we've done. The beautiful thing about it is that so much of this can be replicated easier elsewhere because we've already done a lot of the heavy lifting and actively trained folks how to replicate what we've done.

On the practical side, a tree-planting business can complement a farm operation really well. One of my constant struggles in running Trees For Graziers has been smoothing out seasonal swings in labor, because we have tons to do in the early spring, little in the summer, and variable work in fall and winter. Set up right, a farm operation could fill in some of those gaps quite well. If you already have a farm, then you have the land to run experiments and host farm tours to show off what you've done. You probably have a lot of the same equipment that I've had to buy for plantings—like a truck, ATV, and tractor—and anything you buy for one business can be used on the other as well. Your planting quality for others will be better after practicing and getting feedback on your own place, and your plantings on your own farm will be better after getting experience in all sorts of other contexts.

Don't believe critics, whether around you or in your own head, who say that nobody in your area wants to do this. Unless you live in a desert, there's opportunity for silvopasture (and the value of a few shady trees in the desert is super high, so even that doesn't shut the door). What I've found is that there's a "dormant demand" for silvopasture. If you just hang up your shingle, nobody will come pounding down your door asking for trees. But just about everyone has a place or use for trees—whether that's a broadacre ruminant silvopasture, a corner of the farm or cabin where the landowner likes to hunt, a cider orchard someone has always dreamed about but never made happen, a landowner who needs to install a forested buffer along their creek for conservation purposes, or a backyard chicken enthusiast who wants a grove to produce fruits for her and her chickens. I never would have guessed when I started out that I could build a business of over eight people (in 2025), servicing just our one- to two-hour radius, on silvopasture and stream buffers. And you don't need a big team. This scales down really well to one person running the show, as long as they can get a few hands together for planting days.

If you like the idea of supplementing your farm income with an off-farm job that helps more farmers succeed and build beautiful, profitable, resilient farms, take a good hard look at this opportunity.

THE RIGHT SCALE

In order to create balanced, profitable, and enjoyable operations, we must consider where we act within our strengths and when we start to lose strength by over-extending ourselves and pursuing too much of a good thing. A silvopasture-planting business can be a great fit in your slow season, but if you end up running yourself ragged in every season, something's got to give. Chestnuts and cider apples are sure to find a welcome spot on your table, but you had better have good seasonal labor and an excellent market outlet if you want to move a dozen acres' worth of production. If you have backyard chickens, a handful of mulberry trees will be sufficient. If you decide to plant several hundred mulberries as feed for your cattle, you can add hundreds of chickens to make use of the fruit that the cattle don't eat. That gives you a competitive advantage at the scale of several hundred chickens, but scaling up to several thousand chickens that will require a lot of grain might erode your advantage, making it more sensible and profitable to keep the smaller scale and workload.

The ability to add trees to pastures opens up a world of new ecological and economic niches. Some will be small niches—like when the black locusts you planted for shade now give you flowers that chefs will pay top dollar for. Others will be large, like finishing pigs on chestnuts planted as shade trees. The important thing is to discern which of the many new opportunities we should pursue, when we are working within our strengths, and when we need to pull back or rethink our approach.

CONCLUSION

The thoughtful integration of trees into the farming landscape represents nothing short of a paradigm shift—one that can be applied right away yet will take generations to fully work out. We are so used to the standard logic of farming that to get ahead, you must hold your nose and mine the soil, confine animals, and put the neighbors out of business. It's a crazy agricultural reality when farmers have to work long days doing hard labor for little to no pay, and if they want another operation on the farm, the most common solution is a putrid-smelling hog or poultry house that sinks you into debt in return for a low-wage, degrading job. Nobody gets into farming because it's the best way to make money. They want to farm because of the lifestyle it offers, and they make money to enable the farm life. Yet so many of the industrialized, chemicalized, and commoditized ways that folks have had to adopt to make money on the farm have specifically taken away from that quality of life, so much so that the farm doesn't make much of a life or a living. No wonder so many farm kids have been told to get a career in town.

Silvopasture lets you take the opposite approach. It makes the farm more pleasant to work on, more beautiful to behold, and more attractive for the next generation. The trees insulate your operation from too much or too little heat, too much or too little rain, and the ups and downs of the commodity markets—as well as offering a plethora of ways to stack new enterprises onto the farm without detracting one bit from the existing farm operation. Rather than engaging in a race to the bottom in commodity pricing, you instead

This isn't the farm life anybody wants for themselves or their children.

get to play in a game where peak nutrient density, robust animal welfare, and top-notch customer experiences mean you can fetch a premium for a product that stands head and shoulders above the rest.

These changes will not be fast, of course. But they can happen much faster than you think. I've had many middle-aged clients tell me that they know they won't get any benefits from the trees in their lifetimes—and in three years they are already enjoying shade. The idea that most people overestimate what they can do in one year and underestimate what they can do in ten years certainly applies here. We're used to investing in things that start off nice and inevitably get worse over time, like a barn, a tractor, or a fence post. We're not as comfortable with solutions that start out giving us nothing but get better and better over the course of time. It's not for everyone. But for those willing to build their power of patience, new worlds of opportunity await through the thoughtful, intentional integration of trees into your landscape.

A society grows great when old men plant trees in whose shade they shall never sit.

-Greek proverb

CASE STUDY
Rusted Plowshare Farm and Stacked Ag

By Josh Payne

Around 2015, we fell hard down the soil health rabbit hole, which led us towards a very different vision of our six-hundred-acre farm. Along with my grandpa, Charlie Payne—a very traditional corn/soy farmer—we began an ambitious decade-long transition. What started with cover crops and a continuous living cover annual cropping system quickly moved towards perennial agriculture, both because of the soil health benefits and because of the long-term financial potential and ability to add more jobs on the acres we own.

Because of our location in central Missouri and our proximity to the Center for Agroforestry at the University of Missouri, we naturally gravitated towards chestnuts as a tree crop, establishing a twenty-acre alley cropping and chestnut orchard. Of course we were excited by the potential to sequester carbon and clean water, but we also liked the long-term market research put out by the Center—by year 12 or so, we could expect to harvest two thousand pounds of chestnuts and sell them wholesale at $2.50 a pound. Revenue of $4,500 per acre, with the input cost of corn? And we can continue to farm between the rows? Yes, please! We quickly realized that even a small orchard in row crop country could provide another full-time job, both in work and income.

Unfortunately (or fortunately, depending on your perspective), I found out that my body doesn't eliminate toxins well, and when toxins build up in my system, they cause an anaphylactic response in my throat that could potentially kill me. So we had to pivot out of conventional cropping, fast. We built a bunch of fences and bought a large number of hair sheep, largely because they are more profitable per acre (and lower maintenance) than other livestock species. Though we originally kept them out of the chestnuts, a spell when we were low on grass made us give in and turn the sheep into the orchard. Thus began our love of silvopasture.

We quickly noticed that, if we could protect the trees from livestock while they were getting established, we could cashflow the ground more profitably on a large scale than we could if we cropped or cut hay between tree rows. Not having to take the land out of production to establish the trees made all the difference.

We also noticed that sheep, like so many others, love chestnuts. We use them as a high-energy supplement that drops in the fall, about the time we start to graze winter stockpile. The result? A finishing-quality perennial feed that can be harvested with hooves and rumens instead of tires and diesel. This means I'm not terribly worried about a flooded tree crop market. While not quite so lucrative, grazing chestnuts in the fall provides a nice floor for my perennial tree crop market.

Recently, my brother-in-law decided he wanted to join the farm, so we added pastured hogs as another tree-based enterprise. Like the other enterprises, the hogs have to add at least the value of a full-time job, at wholesale prices, to make this worth our while. Our long-term feed centerpiece is our trees that will drop across three seasons—mulberries, chestnuts, hazelnuts, and thornless honey locusts.

But we wanted to bring my brother-in-law on now, not in five or more years when the trees give significant mast. We'll also keep pigs year-round, so we need year-round feed. Because I don't like doing things the normal way, our primary feed will come from annual grains drilled into a red clover base, strip-grazed, and then healed and reseeded with the next annual grain (oats, wheat, millet, or milo). The result is a short-turnaround, high-value livestock crop while we establish our trees.

Because we had to plant so many trees anyways, we started a small agroforestry service called Stacked Agroforestry, largely with the equipment we use every day on the farm. We have converted inline rippers into tree planters, used our row-crop GPS systems to do layout, and converted sprayers into waterers and truck-mounted grain feeders into mulch spreaders. When they are not in use planting trees in central Missouri, they get used on our farm every day, which has kept startup costs low and allowed us to quickly take on large plantings.

These shade structures work for now, but it'll be so much nicer when trees are providing shade plus feed.

Photo courtesy of Costa Boutsikaris.

SECTION 2
Species Selection

The next step in developing a plan for your farm is to determine which species you will use to accomplish your main goals. We have done our best to break down the options into clear categories: trees for mast (dropped pods, fruits, or nuts), fast-growing trees for shade and fodder, trees for use as windbreaks, and trees for timber. This list is nowhere near exhaustive, and it is focused on species that will thrive in a temperate climate (USDA plant hardiness zones 5–8 with adequate moisture). Yet it represents a solid place to start, and the species listed will be more than adequate for all but the most ambitious and diversity-seeking grazier.

My advice for those not already deeply steeped in trees is to take their time. Start with a small selection of the easier trees to grow and expand from there. Integrating trees into a farm is a process that will not happen overnight. It matters less how fast you go than that you keep going.

Mast trees perform the essential role of providing nutrient and calorie-dense foods for a variety of livestock, whether ruminants or monogastrics. For ruminants, the key use is in supplementing stockpiled winter forages and providing high-energy feed leading up to winter. The feed dropped before winter (apples, acorns, early persimmons) can reduce pressure on forages while grasses are still growing, thus allowing them to be better stockpiled. Feed dropped after dormancy (honey locust pods, late-dropping persimmons) could double or triple the feed calories available for winter use—allowing you to go much longer without hay. For those animals with only one stomach (pigs and poultry) who cannot thrive on forages alone, mast crops offer the easily digested, high-calorie foods they need and usually get from corn, soybeans, or other grains. Whatever the tree and whatever the livestock, in general the goal is for the animals to harvest the food themselves—rather than it being harvested, dried, stored, and then fed to the stock—thus further reducing labor and expenses. The phrase "a dollar saved is a dollar earned" is true if this is a business expense and we're talking about pre-tax dollars. But if you don't write off farming expenses, a dollar saved is actually worth more like $1.33 earned at lower income brackets, and at the higher brackets, a dollar saved is worth two dollars earned! Keep that in mind as you crunch the numbers for the value of these trees.

Photo courtesy of Victoria Moloman, iStock

CHAPTER 5
Honey Locust

While many trees will drop feed, one stands well above the rest for use as winter stockpile—honey locust. Honey locusts with improved genetics deserve a spot on every farm that aims to reduce winter feed costs. The reason is simple—honey locusts drop large amounts of feed late in the season (November through January) that stays good on the ground for months after it's fallen. Additionally, they have very open, dappled canopies, which allow a lot of light through to reach the forages beneath. If one were to pick the most ideal tree for silvo-

This photo of a honey locust research plot at Virginia Tech perfectly illustrates the value of a good silvopasture system. In addition to casting a light shade that moves around throughout the day, honey locusts can bear one hundred pounds of pods per tree per year. A good stand can produce more feed energy than an average acre of Virginia hay ground—at zero annual expense. And there has been zero loss in forage production under these trees. Photo courtesy of Gabriel Pent.

pasture throughout most of the eastern half of the United States, it would almost surely be honey locust. Before anyone gets so upset that they can't read any further, I am very aware of the issue of thorns, and I'll go into more detail about it later. Thankfully, with some genetic selection, that one main barrier is being eliminated. The resulting thornless, female, high-yielding trees should feature in the silvopasture plan for most farmers who have ruminants and can grow honey locust.

It's worth taking the time to dig deeper into the value that honey locust trees can bring to the farm. Everyone understands at a visceral level that shade is valuable, especially if you were to ask during a ninety-degree August pasture walk. That's because everyone has direct experience with being hot in the sun and finding relief in the shade. What most folks are not familiar with, however, are honey locust pods and how much of a game-changer they can be for the way that graziers operate.

Before we get started, though, let's be clear: your results will vary. Trees of the same cultivar tend to give higher-sugar pods the further south you go. Whether you're on rocky soils in New York, a well-drained valley bottom in Georgia, or a windy knob in Kansas will play a big factor in the yields you get. Sorry, but if you're on a hill in Vermont, you'll never get the pod yield a grazier on Mississippi bottomland can expect. So adjust your expectations accordingly.

Let's compare the productivity of honey locust pods to that of grass pasture. I'll use Virginia as an example, because it's a place where honey locusts thrive and is probably the mid-point for pod productivity in the many regions that can support honeys.

According to hay yield data from 2019, average hay yields across Virginia were about 2.2 tons per acre. Let's assume good management and round that number up to 2.5 tons per acre, or 5,000 pounds per acre as feed. Assuming 18 percent moisture in the hay, that turns into 4,100 pounds of dry matter. I realize that actual dry matter production can be significantly lower, especially in a continuous grazing context, but we'll use these numbers for simplicity's sake.

Now on to the honey locust pods. For the sake of this calculation, we'll use one hundred pounds per tree per year as a nice round number. I have commonly seen trees—both wild and from selected stock—produce well over three hundred pounds in a given year. However, honey locusts have a masting tendency where they'll yield heavy one year and light the next. So, heavy and light years need evened out. Thankfully, honey locusts don't sync up like oaks (where all the oaks in the forest yield heavy or light the same years). Therefore, a diverse selection of honey locusts should balance each other out. There is a paper from way back in 1947 reporting the yield from a grove of grafted honey locusts at Auburn University (see appendix). The average yield for the grafted Millwood variety at age nine or ten was 43.5 kilograms, or 95.7 pounds. Being grafted trees, that yield is on the high end for young trees. Unfortunately, we don't have data on how those trees yielded at age twenty or fifty or eighty. So we'll use the round number of 100 pounds per tree per year. Depending on your conditions, the care you give the trees, genetics, and the maturity of the trees, you could get much less or much more than this number.

John Hershey, a leading tree crops pioneer of the early 1900s, wrote in "The Honey Locust: The Hill Sugar Factory," "While in Tennessee in the Tennessee Valley Authority we measured the crop of several large trees. We found 20 mature trees to the acre would produce 5 to 6 tons. Later I learned a forester in the federal forest service had arrived at the same figure." Even if you assume all these trees would take the next year off and cut the figure to two-and-a-half to three tons of pods per acre, that's still a tremendous source of stockpiled fodder, and a low-end production of 250 pounds per tree annually. Keep in mind that only female trees will yield pods, so a clonally propagated orchard will be much more productive than one of seedlings.

This kind of coverage is not uncommon under a mature honey locust tree.

Next, how many trees will this acre have? There is no perfect number of trees to establish, since there are always tradeoffs. Plant two hundred trees per acre, and you'll quickly get to ideal shade levels and good pod production. However, the cost of establishing two hundred high-quality trees on an acre is probably too high for consideration, given the fact you'll need to start thinning them in eight years or less. On the other side of the spectrum, just ten trees per acre will cost less to establish but will take a generation to reach ideal shade conditions and two or three generations before the pod yields equal that of a more densely planted acre. For this calculation, we'll go somewhere in the middle with a spacing of thirty-five by thirty-five feet, giving us thirty-six trees per acre.

Together, the 36 trees per acre and 100 pounds of pods per tree get us to a yield of 3,600 pounds per acre. That's quite the yield boost because it's in addition to the 5,000 pounds per acre of forages, since with such a wide spacing there should be no reduction in forage yields, as we've already seen in the Virginia Tech research. If anything, we would expect to see a slight move of forage production from the spring flush—when we have too much forage anyway—to the summer slump, when forages tend to be low.

We would be missing out on some key insight if we only looked at yield in pounds, assuming that a pound of pods is worth the same as a pound of hay. Saying those two feeds are the same would be like equating a glass of skim milk with a glass of cream. The big difference is that a pound of honey locust pods is much more energy-dense than a pound of hay. So for the sake of a more nuanced comparison, let's look at the caloric energy each feed contains.

Let's start with the hay. At roughly 950 calories per pound of hay dry matter, we're looking at the pasture producing 3,895,000 calories in forages per acre per year.

There's a wide diversity in pod quality among honey locusts. The top pod had only 17 percent sugar, the bottom one had 37 percent, and the middle was likely in the 30-percent range. The two sweetest pods could be cracked open and the sugars squeezed out like jelly.

Now, let's look at what kind of caloric yield comes from those 3,600 pounds of pods. Pods analyzed for the website Feedipedia contained a sugar content of 29 percent and yielded 18.3 MJ/

kg of energy on a dry-matter basis.[14] Translate that to calories per pound for pods, and you get 1,981 calories per pound. This is relatively conservative, given that the pods from the two cultivars studied at Auburn had 37 and 39 percent sugar. We'll assume that the 100 pounds of honey locust pods dropped from our 36 trees contained 12.5 percent moisture, in line with the Auburn research. That then gives us a yield of 87.5 pounds of actual dry matter and a total yield of 173,337 calories per tree, or 6,240,150 calories per acre. Go back and read that again if you wish, and here's a table to show the calculations clearly:

Table 2: Comparison of yield (in pounds) and calorie content of hay and honey locust pods per acre.

	Hay	Honey locust pods
Yield (as fed)	5,000	3,600
Yield (DM*)	4,100	3,150
Calories/lb DM	950	1,981
Calories/acre	**3,895,000**	**6,240,150**
*DM = Dry matter		

This represents a *huge* increase in feed—without buying another square foot of land. It more than doubles the feed yielded per acre, plus shade and all the weight gain benefits that come from that, plus a more diverse and resilient ecosystem, plus a great PR campaign!

Now, this is a simplistic model, as we've already stated. In reality, I don't want people to go create honey locust monocultures. There are a ton of other trees with good practical value in a pasture. If you're reading this, I can be confident you recognize the value of a polyculture. But there's nothing that beats a good honey locust for silvopasture.

While we don't want to plan on honey locust pods or any other mast crop forming the sole or majority diet for livestock, given the high amount of energy, at least in honey locusts the limits have been tested with good success. Christian Dupraz is a French researcher who in 1990 trialed the effects of honey locust pods for sheep feed. Here's what he told me in correspondence, when I asked him about potential dangers of consuming pods:

> Regarding your question of over-feeding honey locust pods to ruminants, I would not be concerned. In our in vivo digestibility experiment we fed rams with only honey locust pods during more than 30 days, and they simply got fat and happy. We did not notice any digestion trouble. At the beginning of the experiment, we milled the pods into a kind of flour, but the sheep refused to eat it. Then we fed the raw pods, and after 2 days of fasting, they had a try and... loved it. Of course we measured that a huge amount of seeds were left intact in the feces, but this was only a small proportion of the ingested seeds. We were trying to find varieties with soft skin seeds, but we did not find any.

Note that in the case of Dupraz, they were looking at pods as a protein supplement, rather than an energy supplement. Different regions have different needs. In the eastern US, cool-season forages tend to have sufficient protein during the winter but are low in energy. The reverse is true in the semiarid West, as well as southern France, South Africa, and other regions where honey locusts have been imported.

Now, let's clear up what is for some the elephant in the room. Not everyone has experience with honey locust thorns, but those that do know them all too well and never fondly. They can pop tires and are a menace to work with. But every indication I've seen tells me the issue can be resolved with some dedicated attention.

[14] V. Heuzé, G. Tran, D. Sauvant, and F. Lebas, "Honey Locust *(Gleditsia triacanthos)*," Feedipedia, 2018. https://www.feedipedia.org/node/295

Not even the most passionate nature lover would hug this tree.

I've been keeping a keen eye on honeys and their thorn expression for years, and it's amazing how much diversity there is. Some regions I've visited, like Wisconsin, seem to have no thorns at all. The South is known for trees sporting extreme thorns, while other farms that I've visited display the whole spectrum—with some trees thornless, some with massive spikes, some with thorns only on the trunk, and some with thorns only on the branches. One particular site had been used for grazing sheep by a previous farmer and featured a small creek flowing through the middle. Because sheep don't love crossing water, the assemblage of trees on the far side of the creek was completely different from that on the near side. The far side had a lush and diverse mix of hardwoods, while the near side, browsed and browsed as it had been, had only cedars and the most gnarly, nasty, thorny honey locusts you've seen. That's all that could survive the constant grazing pressure. Those trees were under intense selection pressure, and only the meanest of the lot could survive. Those were then rewarded because their seeds were later consumed, pooped out, and were that much more likely to grow into well-defended trees themselves. Meanwhile, the same farm had beautifully tall, straight, thornless honeys that would have made great timber, but their offspring never would have held up to constant browsing.

While a grazing operation might accidentally select for thorny trees, we have set out to select for, and then clonally propagate, completely thornless trees. It's been long work, and we wouldn't have done it had it not been necessary.

This Wisconsin honey locust dumped pods and had not a single thorn on it, just like all the others nearby.

Recognizing the value of honeys and the fact that there were hardly any improved varieties available for purchase, we set out to start growing improved varieties ourselves, starting around 2020. The most cost-effective and most scalable way to start was through seedlings. We collected seed from some of the best grafted trees in the country—thornless trees that yielded the biggest, sweetest pods—cleaned out the seed, and worked with nurseries who custom-grew thousands. Some of the resulting small seedlings from the nursery had thorns, but those were weeded out so that only the thornless trees would be sold.

Unfortunately, in late 2022, we started getting some worrying feedback. As some of our initially thornless seedlings were hitting four to five feet tall, more and more were developing thorns—in some cases as high as 75 percent or more of the batch. That was much higher than we had anticipated and a rate that was simply unacceptable for us to continue with.

To complicate the matter, it seems that some batches had been significantly less thorny than others, likely because of the parent trees they were grown from. The trouble is, we hadn't kept batches from different trees strictly separated, so we couldn't go back and isolate which trees yield thorny progeny at a higher rate than others. Because thorns seem to be primarily a genetic trait, certain trees will indeed pass on thorny or thornless traits at a higher rate. What complicates the matter for our selection purposes is two things: all the trees we collected from were visibly thornless, and we don't know which males were pollinating which females.

While trees can be thornless on the outside, we couldn't tell whether they were thornless on the inside. See, you can make a thornless honey locust tree from the most viciously thorny plant. Almost all honeys will stop developing thorns at a certain distance from the ground. Since no living herbivores can reach twenty feet in the air, it's not worth the plant investing in thorns that high up. The height at which honeys stop developing thorns is highly variable, but they do stop at some point. If you want, you can get a cherry picker and take scions (pieces of wood for grafting) from that zone where no more thorns grow. Grafted trees will be thornless if that scion is taken from above the thorn line, even though the genetics inside the tree are still coded for thorniness.

As great of a trick as this might be, its downfall is that the seedlings produced from the tree—whether coming up in a pasture or in a nursery bed—will still be influenced by the thorny genetics of the ortet (the original tree from which a clone came). Because most of the trees at the grove we collect from were likely grafted, and because graft scars are virtually impossible to detect on mature honeys, we had no good way of telling which of the thornless trees we were collecting seed from actually had the genetics for thornlessness and which were simply grafted trees.

In those few instances where we've identified a high-quality wild tree that we know is a seedling and that has no thorns, we would love to collect seed from it but are hampered by the fact that they usually have thorny male trees around that would be contributing their undesirable genetics to the seed. With all this in mind, we determined we needed to pump the brakes on selling seedlings.

Instead, we embarked on a long-term plan to root out thorns from our stock. It started by buying back every single tree that the nurseries had grown for us to sell, culling anything with thorns, and planting the rest at our own nursery to grow out for several years. There they were pruned hard two times, since pruning might induce thorns and we wanted to give these trees every opportunity to show their true colors. Some sources have stated that pruning, browsing, or otherwise damaging trees is the main driver of thorn development, but I have not seen that to be the case. I have seen trees that have been completely cut to the ground, only to resprout without a single thorn. And if thorns were universally developed because of pruning, all of the trees from our nursery would be loaded with spikes. Since that's not the case, I believe that genetics, not damage, are the main driver of thorns. Damage to the tree—like when a branch is cut and the resulting sprout comes back thorny—can indeed cause thorns, but I believe that's only if the underlying genetics are so dispositioned.

After two additional years in the ground and several rounds of culling, we had several thousand trees, ranging from eight feet down to just three feet tall. They were dug and sold, and any excess roots were replanted in nursery beds— a simple means of clonally propagating thornless stock.

Now, there's still no guarantee that those seedlings will continue to be thornless. But I am much, much more confident in a tree that is eight feet tall, three years old, and has been pruned hard multiple times than I am in a tiny seedling just coming out of the field. If the large tree does later produce thorns, I believe they should be small and sparse.

The real prize would be taking root cuttings from trees that are mature and yielding quality pods. The challenge that we've learned the hard way is that their roots lose their viability as the tree gets older. We took root cuttings from half a dozen nice, mature trees, and no shoots grew from them, even though we had solid success taking material from much younger stock. Now that some of our oldest planted seedlings are yielding pods, we can go through, identify the thornless ones with the best pods, and see if their roots will grow. If so, we'll have found the holy grail of honey locusts: inexpensive large-scale propagation of reliably thornless females with really high-quality pods. That is our aim.

I think I can sympathize with ancient dairy producers. The first people that milked an auroch probably got some strange looks, and the casualty rate must have been high. But genetic selection is a powerful thing, and eventually you can turn an auroch into a Holstein. The nice thing about trees is that we can plant thousands and thousands of seeds at once to see what gives us the best results and then clone the best of the best. What took ancient farmers thousands of years, we can accomplish in a much, much shorter time span. I'm sure there will still be lots of room for improvement as more resources are dedicated to advancing the genetics of honey locust. Getting consistently thornless trees that produce pods is the most important first step, and I think we're very close. From there, we can breed and select high-yielders, annual-bearers, trees that drop pods—especially early or late—pods that last longer on the ground, etc. But we are making significant strides already and eliminating the one huge drawback that was keeping honeys from truly contributing to the betterment of farms from Georgia to New England, Texas to North Dakota.

Okay, maybe I hug trees sometimes, but mostly to show off that honey locust trees can be hugged, given the right genetics. This is one of the trees selected by John Hershey for high pod yield and zero thorns, and it has been bearing pods for almost a hundred years at this point.

CHAPTER 6
Persimmons

After honey locust, the next-most-useful tree for winter stockpile is most likely the persimmon. While persimmons can drop from August through February, depending on the variety, it's the late-season droppers that can add to the winter stockpile and are easier to utilize in a rotational grazing context.

A key point to make in the context of dropped feed is that, assuming a rotational grazing system where you're not returning to a particular area for thirty days or more, there's a real question of whether feed that dropped weeks before your livestock pass through will still be edible. While apples or pears or persimmons dropping in August or September may provide valuable energy supplementation and reduce some pressure on forages so they can be stockpiled better, many of those fallen fruits will not still be edible when livestock are rotated back around. As temperatures start to cool towards winter, fallen fruits will last longer. Using the principles of adaptive grazing can allow you to return to certain areas more frequently when fruits are on the ground, as long as they later get adequate rest to recover, but it's certainly more management-intensive than if you were to simply pick up dropped feeds at your own pace over the course of the late fall and winter.

I believe that persimmon can be a real difference-maker for pigs and poultry. Poultry, especially, have very few fruits that they can eat and that drop. While a pig can and will eat just about anything that comes from a tree, chickens have much smaller mouths and nowhere near the force of a hog. We also need to rule out fruits that they could eat, but that don't fall, like serviceberries. Although you could pick and drop them for the birds, that would hardly be an economical solution. After mulberries, persimmons are one of the few dropping fruits that are soft enough to be consumed in quantities by chickens and other poultry. And just imagine the color the bright orange flesh of these carotenoid-rich fruits should give to meat and eggs alike.

Like many fruits, persimmons are high in sugar but very low in protein, so they cannot constitute a full, complete diet. For ruminants, protein tends to be available in abundance in the form of cool-season perennial forages. One combination that might work particularly well is combining high-protein mulberry leaves (which we'll discuss later) with early-season persimmon fruits from late August through October or so, before the leaves drop. Pigs and poultry, which only get a relatively small portion of their diets from forages, will most likely need another protein source—like foraged insects, legumes, or milk waste. If you have dairy processing in your local area, that could provide the protein needed to complement your high-energy tree crops if it's economical to transport and feed.

The bark on an old persimmon is wonderfully unique, looking like an alligator's back.

As I write about using trees as feed crops, half of my work is that of an amateur historian, digging through old books and articles. Most of the relevant information on using crops like honey locust, persimmon, and acorns to feed livestock comes from the days before the Green Revolution and the ubiquity of high-input, get-big-or-get-out agriculture. Since the 1940s, almost no university research or governmental funding has gone into tree crops for livestock feed. Now, however, with more farmers embracing grazing, it's time to dig up that old knowledge about tree crops and make it available to a new generation.

What follows are a selection of quotes from the book Tree Crops, a Permanent Agriculture by J. Russell Smith. Smith was a geologist and fruit explorer who traveled across the globe in the early 1900s studying numerous cultures integrating trees into ag. His goal was to remake agriculture to halt the enormous destruction of the land he was seeing (the book was originally published in 1929, just before the Dust Bowl). The book went on to inspire many generations with its vision of a two-story agriculture made up of grazing beneath and crop-bearing trees above. It is a very worthwhile read indeed, and I know of no better source on the value of persimmons to livestock producers.

The persimmon tree has magnificent natural qualities of great aid as a crop-maker:

- Its extreme catholicity as to soil. It thrives in the white sand of the coastal plain, in the clay of the Piedmont hills and the Blue Ridge, in much of the Mississippi alluvial, and the cherty hills of the Ozark.

- Another soil aspect needs to be emphasized—the ability of the American persimmon to grow in poor soil. I have seen them grow and produce fruit in the raw subsoil clay of Carolina roadsides and in the bald places in the hilly cotton fields where all the topsoil has been washed away.

- The persimmon is remarkable in the length of its fruiting season. With the persimmon, nature unaided has rivaled the careful results of man with the peach and apple, for the wild persimmon ripens often in the same locality continuously from August or September until February, dropping their fruits where animals can go and pick it up through this long season of automatic feeding. In this respect it is ahead of the mulberry. Furthermore, it should be pointed out at once that this long season combines the added virtue, the great virtue, of. . .

- Automatic storage. It is true that nearly half of the season of persimmon-dropping occurs after frost has stopped all growth, and farm beasts are usually eating food stores in barns. Truly these are two great virtues for a tree crop.

- The fruit of the persimmon is very nutritious. It is said to be the most nutritious fruit grown in the eastern United States.

- It is too much to expect the persimmon tree to have the complete and amazing collection of attributes cited for the mulberry. Compared to the mulberry, the persimmon (a deep rooter) is not easy to transplant. Therefore it is produced in the nursery at greater expense. It does not grow so rapidly as the mulberry. It does not even grow quite so rapidly as the apple. However, it is a much more common tree in America than the mulberry, growing wild in much greater abundance. This is partly due to the fact that the leaves are shunned by most pasturing animals, including the sheep and goat. I have proved this in my own experience. This is a point of great significance because seed can be planted in pasture fields, pasturing can continue without interruption, and the trees can be grafted to better varieties once they reach a suitable size.

- Native persimmons in my native locality, northern Virginia, bloom late in June, when wheat is ripe. This almost sure escape from frost injury seems to be a great advantage.[15]

[15] Smith, *Tree Crops*, 99.

Smith goes on to share from conversations and correspondence he's had with numerous farmers and specialists over the years, giving us a view of how widely persimmons were used throughout so much of the country in that era:

The beauty of persimmons compared to many other fruit trees we're more familiar with is that they are very low maintenance once they get established. With no pruning, spraying, or real care, they can still yield bushels and bushels of fruit every year.

> "The seedling persimmons we have grown here have in the fifteen years they have been growing attained a height of from twenty-five to thirty feet and a diameter averaging about six and a half inches. These trees have been growing in good ground and have been given good care. . . . We consider from four to five bushels [180–225 pounds] a fairly good yield for these trees." (Albert Dickens of Kansas State Horticultural College, 1916)

"In North Carolina I have seen boys getting persimmon from the trees where the hogs were pasturing, and it required three boys to get the persimmons—one to keep the hogs back, one to knock the fruit from the tree, and one to pick it up. And the boys had to be quick if the hogs did not get a share." (D.S. Harris, 1913)

"Here in southern Indiana where native persimmons abound, nearly everyone appreciates the value of persimmons for hogs. I had a hog pasture on one of my farms in a native persimmon orchard, and their value is almost equal to corn, after they get ripe, but cannot be utilized at any stage of growth like corn. But after they begin to ripen and lose their astringency, they turn rapidly to sugar that has a decided fattening property." (Alvia G. Gray, 1913)

These beautiful fruits are truly loved by all—humans, cattle, pigs, and more.

"It is very common in the making of a hog pasture to retain any native persimmons that may be growing there with the idea of the hogs gathering the fruit." (C.C. Newman of Clemson College)[16]

It's worth extrapolating those figures for persimmon yield to a per-acre basis to give us a better sense of the yield. Albert Dickens writes from Kansas—which is on the far western range of where persimmons will grow—and says they consider four to five bushels of persimmons per tree to be a good yield.[17] Planted at thirty-by-thirty-foot spacing yields forty-eight trees and 240 bushels per acre at the five-bushel-per-tree mark. If we assume that a bushel of persimmons weighs 56 pounds, that's 13,440 pounds of fresh fruit per acre, which lines up well with industry reports measuring 6.95 tons of Asian persimmons per acre for human consumption. That compares favorably with honey locust—on a tree that tends to grow smaller and narrower, with a much smaller crown.

Ripe persimmons are universally appreciated by livestock and hence form a bridge between ruminant and monogastric systems. Late droppers that can stay on the ground for extended periods can do service for ruminants as they make their way around the farm, while early droppers might make the most sense clumped into groves, staged in such a way as to make it easy for pigs, turkeys, and chickens to harvest. That is where we'll turn our attention next.

---

16 Smith, *Tree Crops*, 101–103.

17 "Persimmon," Agricultural Marketing Resource Center, April 2024. https://www.agmrc.org/commodities-products/fruits/persimmon

CHAPTER 7
Mulberries

While honey locust should be considered the king of stockpiled feeds for ruminants, I believe that mulberries are the most valuable tree when it comes to feeding monogastric livestock—and it's not a close competition. That wouldn't be the case if mulberries dropped their fruits in the fall, when there's a true glut of feed coming from apples, pears, persimmons, oaks, chestnuts, hickories, and more. Rather, the fruits drop early in the summer, during a time when almost nothing else is falling to the ground except for maybe some plums, June drop apples, or goumi berries. That's about it. As such, they fill a critical niche and can offer months' worth of valuable, high-energy fruits.

Fascinatingly, mulberry fruits also have a surprisingly high protein content. While their leaves have long been studied on account of their high protein content and digestibility, the fruits have received very little attention when it comes to feed analysis. So I decided to see what I could find. I started off by going down an internet rabbit hole, working mostly off of obscure Chinese studies and pieced together a picture that looked like mulberry fruits could have protein contents of 11 percent or more—up to 28.7 percent for one cultivar. And the amino acid profiles looked very strong, similar to or greater than corn for those aminos that are most limiting for livestock growth. Because I wasn't sure I had done the numbers right or had the right sources, I decided to do my own testing. Once June came around, I harvested from two of my young trees that were yielding the highest, sent the samples in for testing and, indeed, got very significant protein results. The cultivar known as Oscar yielded 12 percent protein, while Shangri-La came in at a whopping 18 percent protein. Again, these are fruits I am talking about, not leaves, which is what made it so surprising. In comparison, apples have 5 percent protein on a dry matter basis, and American persimmons have just 2.5 percent.

The amino acid profile of the fruits was lower than expected given the high crude protein content, but the values are still on par with corn, and crude protein for Shangri-La fruit is double that of corn. Interestingly, where mulberry fruits are below the amino acid nutritional requirement for pigs is in lysine and threonine, both of which are high in mulberry leaves. Thus, the tree is uniquely suited to offer a complete ration, and mulberry leaves are routinely used as a fodder supplement for hogs in certain countries due to their high protein and digestibility. I am no researcher, but I am hoping this will get picked up by researchers with time, budgets, and a knack for details, because it seems like there are tons of nuances to pull out. For example, what factors influence the protein and amino acid content? Cultivar is one, for sure, but I bet there are differences based on soil type and whether trees are being pollinated and setting seeds. It would be amazing to find cultivars that routinely provided very high protein with even higher rates of the most limiting amino acids.

It's worth noting that Shangri-La fruits were selected for testing because they are very large, easy to collect, and yield very heavily. The flavor is not great, but the nutritional profile turned out to be very strong. Oscar, on the other hand, is one of my favorite-tasting mulberries and also yielded heavily. The fruits, however, are only about a third the size of Shangri-La fruits, which is a factor to consider for livestock use. Chickens likely don't care too much, and in fact they can much easier swallow a fruit whole when it's small. But pigs are likely to prefer larger fruits, and if you want sheep, goats, or cattle to pick them up for an energy boost, you'll want the biggest fruits you can find.

Cultivars like this drop fruit like it's their job. This one dropped really heavily for about three weeks, then lightly for another week or so before being done for the season.

Besides selecting the best cultivars and making sure there are trees around to pollinate, there's another interesting workaround for increasing protein: bugs. Insects of all kinds—from ants to fly larvae—like mulberry fruits, and the more bugs, the higher the protein of whatever a chicken or pig ends up eating. I'm curious if a system designed such that the animals get access to a section of fallen fruit every several days, with a few days for fruit to drop and grubs to grow, might result in higher levels of protein consumption than if the fruits were all eaten fresh off the tree. Again, there's no shortage of work here for a small army of grad students.

It would be enough if mulberries dropped their fruits early and had a high fruit protein content. But they also yield prolific crops of fruit. Again, there is no better resource on the subject than our friend J. Russell Smith, with statements such as the following:

> I find a very general belief in the Cotton Belt that one "everbearing" mulberry tree is enough to support one pig (presumably a spring pig) during the fruiting season of two months or more. Professor J. C. C. Price, Horticulturist, Agricultural and Mechanical College, Mississippi, says, "The everbearing varieties will continue to bear from early May to late July, a period of nearly three months. I believe that a single tree would support two hogs weighing 100 pounds each and keep them in a thrifty condition for the time that they are producing fruit. They could be planted about 35 trees to the acre."[18]

Smith also shares this colorful account, my favorite of the book:

[18] Smith, *Tree Crops*, 86.

> In that east-central part of North Carolina where the mulberry orchard is a very common part of farm equipment, a veteran of the Civil War (a captain) declared, "When I lived ovah the rivah we had a lot of mulberry trees—300 to 400 mammoth big ones. We had fully 200 hawgs, but we had to send fer the neighbors' hawgs to help out and to keep the mulberries from smellin'." Where the captain then lived he had a bunch of thirty-five hogs of various sizes running in mulberry and persimmon pasture, all of which I saw. He estimated that one-third of their weight, or one thousand pounds of pork, live weight, was due to the mulberries from eighty trees set twenty-four by thirty-three feet. That runs out about six hundred and twenty-five pounds of pork, live weight, to an acre of rather thin, sandy land with little care and no cultivation. A big yarn, you say? I'll willingly take it back just as soon as any experiment station makes a real test and disproves it.[19]

Mulberries can yield a ton. The challenge is having farm systems built around the seasonal use of a very perishable feed that livestock must collect themselves. For those who build their farms that way, they will have free access to a tremendous amount of high-quality feed.

The challenge with this type of anecdotal evidence not figuring out whether it makes sense to plant these trees. The fact that so many folks planted them in the day, without government cost-share or premiums for mulberry-finished, pastured pork, speaks volumes. There is zero question about whether these trees will contribute to a farm. It's more about figuring out how much they are worth to your farm. They might be great to have, but is it worth investing in mulberry trees versus the hundred other things you could put your money towards?

It's really interesting to crunch the numbers and figure out just how valuable these trees are. Feed costs are, after all, the largest variable expense for raising hogs. According to research in North Carolina, feed costs made up 45–63 percent of the cost to raise a hog, depending on the cost of feed.[20] So, anything that can be done to reduce feed costs should be seriously considered. Let's run the numbers and find out just how valuable mulberries can be.

Here are the assumptions for my calculations:

- The hogs eating mulberries are one hundred pounds at the beginning. That's the assumption made by two folks quoted in Tree Crops, so we'll go with those numbers.

[19] Smith, *Tree Crops*, 87.

[20] D. Washburn, "NC Farm School Pork Info-graphic Breakout," NC Farm School Program, North Carolina State University, 2016. Retrieved from https://ncfarmschool.ces.ncsu.edu/wp-content/uploads/2016/08/NC-Choices-NC-Farm-School-Pork-Info-graphic-Breakout.pdf?fwd=no.

- We'll give versions for both thirty and ninety days of mulberry yield. The farther south you are, the longer mulberries tend to yield. It also seems that the more nutrients (like hog or chicken manure) trees have access to, the longer they'll yield. Many mulberry trees will only yield for about thirty days, so that is likely a more realistic version for folks in USDA plant hardiness zone 7 or colder. That said, many mulberries that drop over a shorter period drop heavier amounts at once than everbearing types, so they would likely offset a greater portion of other feeds for that shorter period.

- We'll assume that mulberries will reduce feed bills by 66 percent, not the 75 or 100 percent that some mention, as a means of being more conservative with our assumptions.

- We'll use three different prices originally used in the study from North Carolina, broken down by organic, non-GMO, and conventional. You can decide which best fits your production system.

- We'll assume that hogs are eating five pounds per day on average over this period.

The results are pretty impressive, as you can see from the chart. On the very conservative end—with hogs on conventional feed supplementing their diet with mulberries over just thirty days—we get a total feed savings of $21.78 per hog per year. On the high end—with organically fed hogs supplementing their diet for ninety days—the savings can amount to $136.62 per year.

This everbearing mulberry is still working on new fruit as of mid-July, several weeks after most other trees in our area have stopped producing. It doesn't drop nearly as heavily all at once, but the long fruiting period can have real benefits.

Table 3: Savings in feed cost for hogs foraging on mulberries for 30 or 90 days, based on costs for organic, non-GMO, and conventional grain.

Feed Type	Days	Pounds of feed saved*	Feed cost per lb	Feed savings per hog	Value per acre†	Upper limit of investment over 5 years‡	Upper limit of investment over 10 years‡
Organic	30	99	$0.46	$45.54	$1,639.44	$268.91	$158.79
Organic	90	297	$0.46	$136.62	$4,918.32	$806.73	$476.36
Non-GMO	30	99	$0.37	$36.63	$1,318.68	$216.30	$127.72
Non-GMO	90	297	$0.37	$109.89	$3,956.04	$648.89	$383.16
Conventional	30	99	$0.22	$21.78	$784.08	$128.61	$75.94
Conventional	90	297	$0.22	$65.34	$2,352.24	$385.83	$227.83

*Per hog, at a feed rate of 3.3 pounds per day
†Based on one tree per hog and 36 trees per acre
‡Per acre, at 10% discount rate

There's something groundbreaking hiding in the numbers above, something that took me way too long to figure out. Let's split the difference between the above scenarios and estimate that a tree would offset 198 pounds of grain. Now, how much did that cost us to produce? In order for it to be worth investing in, it should cost less than what you're paying for similar feed.

Let's assume the cost of establishing that tree was $50 and you got a twenty-year loan at 6 percent interest to cashflow the purchase, giving you a total repayment amount of $85.92, or $4.30 annually. Let's also recognize that for the first several years you have little to no yield, and that you only get fifteen years' worth of yield in the first twenty years, or 2,970 pounds. That gives you a grand total of $0.03 per pound of feed over those twenty years. Read that again, and no, I didn't put the decimal point in the wrong spot. That's not 30 cents, but rather 3 cents per pound. That's less than one-tenth the cost of the non-GMO feed.

In a world where pastured pig and poultry producers are paying several times what their confinement counterparts pay for feed, 3 cents per pound is nothing short of revolutionary and completely flips the script. This means the cost of feed can be significantly less for pasture-based producers than even CAFO operations—for feed that is organic, non-GMO, corn-free and soy-free! It's the most premium feed at a fraction of the cost of the cheapest feed available! And after the loan term, the rest is completely free. It's an investment in critical farm infrastructure that, rather than deteriorating like a dust-caked poultry barn, only gets better and better for generations to come. You can do the math for other tree crops like persimmons and chestnuts and find that they also come out costing significantly less than grain inputs, even if they can't quite match the volume and earliness of production achieved by mulberries.

Now, not all mulberries are created equal, and not all will perform in a stellar way on a farm. Seedling mulberries will either be male or female, and only female trees yield fruit. Of the females, not all yield heavily, produce over a long period of time, or have especially tasty fruit. I will be very interested to learn whether there's a correlation between fruits that are sweet and tasty to us and those that have the best nutritional value for livestock. In some cases, you might want everbearing trees that drop their fruits gradually over many months, but other contexts would prefer trees that drop buckets full of fruit over a few weeks, after which you move on elsewhere. The beauty of using clonal stock is that you know that you'll get a very high-quality tree that yields well and you'll also know how long it will yield and how it fits with your other trees.

So far we've mostly focused on the fruit yield of mulberries, but it's worth appreciating how valuable their leaves are as well. They are well-known for being high in protein and very digestible. All types of livestock readily go after their leaves, and the trees are very adapted to growing back from browse. They could be used either as a standard tree where the main purpose is to provide shade and fruits—and limbs in event of a drought—or they could be maintained as a

I caught my neighbor's cows in the act of browsing on several mulberries, including a grafted tree. Thankfully, mulberries are accustomed to abuse and come back vigorously.

shrub, with livestock having access to browse the trees so that you don't need to cut limbs down. I believe that dwarfing mulberries, which don't get taller than about twelve feet, should be ideal for this application, because they never develop a large trunk and could be cut back with a brush hog or forestry mulcher if cut regularly. That type of use would have mulberries yielding fruit, leaves, shade, and even windbreak. That's a tough combination to beat.

While honey locusts may have the widest applicability in silvopasture and anyone grazing ruminants would do well to plant honeys across much of their farm, I believe that mulberries have the highest return on investment for any stock-feed tree. Much of that comes from the fact that, rather than offsetting relatively inexpensive hay, you're offsetting grain in the diet of pigs and poultry. I believe that mulberry trees are key to seeing more grass farms enjoy strong profitability by producing some of the best meats the world has ever seen while keeping costs very low.

These birds were truly living their best lives under a mulberry tree.

CHAPTER 8
Other Masting Species

Now that we've covered the top two trees for ruminants (honey locust and persimmons) and the top species for monogastrics (mulberries), I'll wrap up the section on masting species by discussing other species with more limited-use cases in silvopasture—apples, pears, oaks, and chestnuts. These make up the cornucopia of tree crops that drop during the fall, having stored up solar energy in the form of fruits and nuts. Most are primarily feed for hogs and poultry, although there's real room for ruminants to eat them as well, though probably in smaller quantities.

There's a lot to like about apples and pears for use in silvopasture. Both are a great source of energy—which is a main limiting factor for ruminants on perennial pasture—and both have hundreds or thousands of cultivars to choose from. Some drop as early as July, while others will hold on until deep in the winter. I believe that the most robust, late-dropping storage apples likely have a strong use case as a stockpiled feed, similar to persimmons.

There is some controversy around how cattle react to apples, and this will likely translate to other high-energy tree crops like persimmons. Any farmer knows that when transitioning livestock from one feed to another, it's important to introduce the new feed gradually, rather than suddenly switch from one feed to a very different type. If ruminants eat too many sweet apples or pears or persimmons or anything else all at once, they are susceptible to acidosis, a condition that can be fatal. That's why many farmers have been advised to keep their cattle away from apple trees.

The challenging thing is that we know that a moderate amount of apples are not a problem at all when introduced gradually and can in fact induce greater yield and weight gain. I can look out my office window and see a pasture where every year two steers are finished. They have constant access to apple trees, chestnuts, and oaks, and they spend a lot of time under those trees when they are dropping fruit, all without issue.

The best example of a farm intentionally using apples for cattle is Walnut Hill Farm in northwestern Pennsylvania, run by Michael Kovach. He has hundreds of seedling apple trees, which may or may not have been planted by Johnny Appleseed himself, depending on who you believe. Regardless of who planted them, the cattle love them, and Michael moves his cattle so that they get regular access to the apples during the drop period. Knowing the value in the apples, Michael practices adaptive grazing by bringing the cattle to the apples for short periods so they can pick up the fruit before returning to other pastures. Sure, the grasses under the trees will need a long recovery period after being grazed every few days, but it's a worthwhile tradeoff for Kovach in order to capitalize on the high energy their diet is otherwise lacking. He reports no serious issues with this practice in the many years he's been doing it.

I am no veterinarian, so I cannot say what's best for you or your animals. But this is worth exploring. There are certainly dangers to be weighed and considered, but there are also real upsides to be gained through the thoughtful, cautious integration of trees like apples and pears.

Oaks have a similarly mixed reputation among graziers, and various well-respected folks have very different perspectives on them. Their status as a quality feed for hogs is legendary—farmers have relied on oak forests to feed pigs since well before Roman times, and Rowan Reid, a leading agroforester from Australia, notes that he has to compete with his sheep to get a share of the English Oak acorns if he wants to plant more—he calls them valuable fodder trees. Yet many folks have heard horror stories about cows keeling over or drying off prematurely from overfeeding on acorns.

Eric Tonesmeier, a very well-respected authority on tree crops, writes the following in his paper on oak domestication:

> While some people (including the great George Washington Carver) recommend acorns for cattle and small ruminants, quite a few sheep and cows are killed every year by the gallotoxins in acorns (especially unripe acorns), so I can't recommend that for the moment. Lucky Pittman is not only an oak expert but also a diagnostic veterinary pathologist. He states: "Whitetail deer and goats possess proline-rich salivary proteins which bind and inactivate tannins, allowing them to consume acorns and browse high tannin leaves and forages with impunity. Cattle and sheep lack these proline-rich salivary proteins, and as a result 'acorn toxicosis' can be a significant concern during 'on' mast years, if these animals have access to large amounts of acorns. During my early years in veterinary practice, and the latter 30 years or so of my career as a diagnostic veterinary pathologist, I saw numerous cases of acorn toxicosis in cattle."[21]

Like so many things in livestock management, the key seems to be how much was fed and how much time there was to acclimate. If a cow goes from an all-grass diet to all acorns the next day because the grass runs out, expect trouble. In smaller, more reasonable quantities, cattle tend to like acorns and oak leaves, often going for those first thing when they enter a new paddock. As the saying goes, everything in moderation.

One often hears that you should plant white oaks instead of red oaks, because red oak acorns are higher in tannins and hence less palatable. This is often highlighted by hunters looking to lure in a trophy buck at just the right time of year. But there are real reasons to consider planting red oaks as well, whether for deer or cattle. White oak acorns sprout very soon after hitting the ground, trying to develop a tap root before winter sets in. Red oaks take a different strategy and wait until spring to sprout. Hence, they tend to have higher levels of tannins to protect the acorn over the winter, from both predators and decay. What that means for cattle is that acorns higher in tannins tend to last longer on the ground, making them an interesting addition to a late-winter stockpile diet. Red oaks also tend to stay on the tree longer, where they are even less likely to rot. While they may never be a staple crop for your silvopasture system, both white and red oaks can make an interesting addition to a diverse system while also providing ample food for wildlife and just being majestic, long-lived trees that will produce beautiful wood when their days as stock-trees are over.

The significant challenge of establishing

[21] Eric Toensmeier, "The Status of Oak Breeding and Domestication as Food for People and Livestock: In South Korea, Spain, and the United States," *Perennial Solutions*, 2023. Retrieved from https://www.perennialsolutions.org/wp-content/uploads/2023/09/Status-of-Oak-Breeding-2.0.pdf.

oaks is that they are tap-rooted and do not appreciate being transplanted. In the time it takes a willow or black locust to get over transplant shock and make six feet of growth, an oak might only be starting to gain a couple of inches. This is challenging in a pasture setting, where size matters and there are a lot of animals who would love nothing more than to eat the tree. The same limitation applies to hickories (mainly hog feed) and persimmons. To overcome this limitation and get these trees up to size fast in a silvopasture, I like to plant these trees rather tall, ideally over three feet. While they might experience more transplant shock than a tiny eight-inch seedling that you get from a state nursery, the head start seems to more than make up for the difference when they've been given a good heavy dosage of mulch around their roots to soothe the transplanting process.

I would love to see more of these majestic, ecological keystone species in the landscape. Their best use is most likely for pigs and turkeys, with some acorns being consumed by cattle and some of the species with smaller nuts—like the willow oak—potentially being useful for chickens. They will probably never be as profitable as a honey locust for ruminants or a mulberry for pigs and poultry, but for those with particular patience, they will be long-lived and productive monuments on the farm.

For more on the subject, see George Washington Carver's classic pamphlet titled *Feeding Acorns*.

If you are looking for a tree that can provide an abundance of food for both humans and livestock, chestnuts should be your first consideration. Their ability to turn sunlight into sweet, tasty nuts is unmatched in temperate climates, and everything—from sheep and cattle to pigs and turkeys—appreciates their flavor. I have observed cattle in a pasture where both chestnuts and apples were dropping, and they spent significantly more time under the chestnuts, waiting for the next one to fall.

These willow oak acorns are small enough that they shouldn't pose a problem for poultry, and their nut meat has a rich color.

One of the reasons we have done very little work with chestnuts in our silvopasture plantings to date is that their nuts are high in starches, similar to corn, and hence are more likely to cause issues with folks who run certified grassfed operations or plan to do so in the future. While honey locust, persimmons, and apples provide energy in the form of sugars—which are permitted in grassfed programs and are often supplemented through molasses—I estimate that chestnuts are more likely to raise red flags from certifiers if present in large quantities. You will need to be the judge of whether that makes a difference for you. They also don't keep as well on the ground as a honey locust pod, though they should last considerably longer than an apple or persimmon dropped at the same time.

Given that they are relatively easy to establish and seedlings with quality genetics are easy enough to find at scale for reasonable prices, this multifunctional tree should see considerable use—especially on operations that run multiple livestock types or who are interested in the human food aspect.

CHAPTER 9
Shade and Fodder Trees

This category includes those trees that serve two main functions on the farm: fast shade and summer fodder. These trees are easy to establish, grow fast, are immensely forgiving of abuse, and, if planted in volume, can add significant amounts of feed during the summer, either during the typical annual summer slump or a more extreme drought.

SHADE TREES

Introducing shade to pastures is the number one thing that brings most folks to silvopasture, and most want it yesterday. That's why we turn to certain species that reliably grow very fast and are inexpensive to establish. While they might not be the trees that will contribute the most in the long term because of their lack of mast to feed livestock, they will pay for themselves by getting to size faster and providing much-needed shade.

Willows and poplars have the rare superpower of being easy to propagate via dormant cuttings. You just take a dormant branch and push it into the soil, and it will take off from there to establish its own roots and leaves. If all trees were that easy to grow, we would have silvopasture everywhere. We have taken this practice to the next level by using very tall and large live stakes, instead of the twelve-inch stakes that many conservation groups are familiar with using. Ten feet has become our standard height, although we have planted stakes in excess of twenty feet, mostly to see if they could work. So far we have had very good success with those extremely tall live stakes.

This opens up the possibility of what I call the Traveling Tree Trick. If you're interested in getting shade spread all around your farm but are constrained by your budget, I recommend starting with willow and poplar poles first. Let's say you start with ten-foot poles in some of your highest-priority areas where you really want some relief from the heat. After several years, once there is enough shade established, you can cut some of them during winter and simply plant them elsewhere. If you cut them above browse height, then you get a two-for-one where the original stays and can resprout. If you cut the original to the ground, it will provide browse when leaves reemerge. If the tree trunk is large enough and you can successfully auger a hole about three feet deep, you can likely get that new tree established without a tree shelter, because the large pole with hardened bark will be sufficiently strong to withstand some rubbing. It's still best to protect it for the first year, but this is an intriguing possibility for establishing shade trees in areas that are too challenging to use fencing.

While willows and poplars could aptly be described as fast-growing yet short-lived, black locust and sycamore can live much longer—black locust can live over fifty years, and sycamores commonly live for several hundred years. We have typically used black locust mostly as a

This live stake was taken from a dormant twenty-foot hybrid poplar. The tree was cut, side branches were removed, and the stake was packed into a three-foot-deep hole in moist ground. Given the sturdy width of the trunk, we chose not to protect this or the others planted nearby, and most survived the first year in good condition.

quick-growing shade tree, with the intention of removing it once the shade is no longer needed and using the trunk as fence posts or firewood. Yet the tree could also be used at lower densities for long-term shade, nitrogen fixation, and pollinator forage, since beekeepers love the species. Mature black locusts tend to be tall with few lower branches—an ideal shade profile for pastures.

Sycamores are one type of tree that I've only recently come to see real potential in. They certainly will never give a high economic output for a farm, because they don't drop fruits or nuts or pods, likely aren't a great fodder species, don't fix nitrogen, and don't produce a valuable saw log. So why would someone want to plant them? Because they are majestic. And sometimes that's what matters more than the economic yields. The stunning white bark of a mature specimen can light up a winter landscape—a beautiful landmark for miles around. It doesn't hurt that they are also fast-growing and easy to establish, even via tall live stakes. They have tough bark that holds up well to rubbing and have very tall canopies that move shade far throughout a day, so just a small handful of larger trees per acre can provide all the shade that's needed for a farm and distribute it enough to avoid livestock camping in any one space. It's not one I would recommend for most graziers to use heavily. But so many landowners just want some trees that are simple to establish, look nice, and don't cost too much—and for that, sycamores check all the boxes.

I tested whether we could get sycamore live stakes established in an actively grazed pasture without protection. Results were mixed, and sycamores seem to live stake less well than willows or poplars. We will test whether pre-soaking in a rooting hormone will lead to better success rates.

There's a reason why there's a pretty strong overlap between fast-growing shade trees and those we recommend for fodder—fodder species need the vigor to recover from being browsed. That's likely a mechanism that's baked into the tree's DNA. While some species invest heavily in making themselves less palatable, others make palatable leaves they know will be eaten and just make up for it by quickly overcoming the bite and growing out of reach. Those are the natural traits we want to harness for livestock feed.

FODDER

Leaf fodder from these trees can be made available to livestock in two main ways— pollarding or coppicing. Pollarding is the act of cutting a tree above browse height so that it regrows from that spot. Each tree can be pollarded on a one- to three-year cycle, and the act of cutting it back actually extends the lifespan of the tree by keeping it in an actively growing, juvenile state—much like keeping forages in a vegetative state and preventing them from going to seed. Coppicing is the act of cutting a tree or shrub down to ground level and allowing it to resprout, usually with many more shoots than it started with. It looks like a pollard, only at ground level.

An old pollarded willow in Belgium, where pollarding has a very long tradition. The height of the cut is likely on account of the tall Belgian horses that often share the pasture. Author's father for scale.

There are some key differences between these two methods when it comes to managing them for fodder in a grazing operation. Pollarding has the distinct benefit of being able to tightly control when the leaves are available for livestock to eat, since animals cannot reach the leaves unless you cut the branches down for them. It does, however, require labor to cut those limbs down and clean them up, if you so desire. That is where coppicing becomes interesting, especially for larger-scale applications. Coppicing brings leaves to the browse height of the livestock, meaning they can harvest the leaves themselves. It does require that you protect the shrubs from browse anytime you don't want them to be eaten, especially since shrubs cannot handle being browsed as frequently as forages can. Another benefit of thick, coppiced rows or hedges is the windbreak factor. To coppice a row of trees, simply cut it back to near-ground level every year or two or three. A forestry mulcher could do the job without a problem, but ideally you'll be able to keep at it regularly enough that you will never need more than a brush mower.

Jaime Elizondo, a grazier, consultant, and silvopasture practitioner in Florida, notes that he has decided to use pollarding over coppicing in his silvopasture systems (featuring the sub-tropical leucaena) because he can control livestock access better with pollarding and was experiencing higher rates of damage and mortality when trees were coppiced and browsed. He mentions that al-

Shrubs that stay small and don't get a large trunk, like this Streamco willow, are an interesting option for coppice systems because they can be readily managed mechanically.

though it does represent some labor, he is able to feed his entire herd of 250 cows with only two hours per day of pollarding.[22]

The main factor in determining which of these species to use might be the soil conditions where you want to plant them. Good luck getting a black locust to grow in a swamp—yet that's exactly where a willow will thrive.

There are, of course, other trees that can be used for summer fodder. Species like elm, ash, and linden have been used traditionally and have high feed values. If you have these on your property already or want to add more diversity in your browse system, great. For simplicity's sake, we will focus on those species that are most widely used and easiest to establish.

While the most obvious benefit of tree fodder is that you can use it to get through the summer months, there are wider-ranging benefits as well. For those in areas dominated by endophyte-infected fescue, systematically establishing quality tree fodder species can give your livestock a more useful option than fescue. Reducing heat stress with shade and a less fescue-heavy diet is a double win. Fescue pastures are notoriously challenging to renovate into new forage stands, and I believe that, with the right site prep, establishing browse shrubs could be a simpler approach. Dwarf mulberry cultivars—which never grow much above browse height and have high-protein, ultra-digestible leaves—are a great option.

Another avenue that needs experimentation is the use of high-tannin fodders specifically to offset the effects of fescue. Dr. Allen Williams, a founder of Understanding Ag, said the following when sharing strategies for mitigating the harmful effects of endophyte-infected fescue:

> First—it helps tremendously when you are able to build the mycorrhizal fungi population in your soils. We have found that the infectivity level goes way down when you have a strong mycorrhizal presence. The mycorrhizal fungi counters the endophyte.
>
> Second—promoting and facilitating plant species diversity is crucial. More diversity dilutes the amount of alkaloids the cattle are consuming from the endophyte-infected fescue.
>
> Third—including in that enhanced diversity plants that contain tannins. The work of Dr. Fred Provenza shows that tannins and alkaloids essentially cancel each other out. If you have a good mix, cattle can eat a lot more of both the alkaloids and the tannins and it benefits them.

22 "How Leucaena Can Double Cattle Units Per Hectare and Transform Profitability with Jaime Elizondo." The *Regenerative Agroforestry Podcast*, episode 28, September 15, 2021, https://feeds.acast.com/public/shows/regenerativeagroforestry.

> Tannins are a class of astringent, polyphenolic compounds that bind to and precipitate proteins, including amino acids and alkaloids. Tannins not only temper the effects of the alkaloids from endophyte-infected fescue, but they also are natural dewormers and help prevent bloating.

Fodder tree species can address each of those three because many are high in tannins (especially willow and black locust), they add to the diversity of forages available, and they are powerful means of enhancing mycorrhizal communities in a pasture because trees have a strong mutualism with mycorrhizal fungi.

Mycorrhizal fungi are trading partners with plants. The fungi are great at gathering water and minerals, and the plants are great at turning sunlight into energy. Like flowers and bees or zebras and oxpeckers, it's a mutualistic relationship where both parties benefit. The major reason tall fescue outcompetes other forages is because of a symbiotic relationship with an internal fungus that makes the plant toxic to would-be herbivores. While herbivores happily chow down on other plants, the fescue is among the last to be eaten, which allows it to put more energy into its roots and hence tolerate heat, drought, and stress much more than it could without that partnership. Developing a strong mycorrhizal community levels the playing field by giving more support systems to plants other than tall fescue. Combine that with more tannins that allow cattle to tolerate and consume more fescue—as well as shade that cools the other cool-season forages—and you have a combination that could be a real game-changer for the roughly forty million acres currently dominated by fescue.

These fodder species will bear leaves with very different nutritional profiles from the standard forages available in your pasture, which adds a richness of minerals for livestock to choose from. Steve Gabriel, author of the book Silvopasture, published forage testing results comparing the makeup of six species of trees, leguminous pasture, and grass pasture. Willow stood out for high content of calcium, zinc, and manganese, while black locust was strong in copper and manganese. These species had relative feed values of 200 and 227, respectively, which were both higher than the tested legume pasture (192) and grass pasture (98).[23]

Even for those who do not regularly experience a summer slump, tree fodder should still be considered because it can help with building your winter stockpile. When you get to the point of the year when you start building up a stockpile, having tree fodder to feed your livestock can reduce the pressure on your forages, allowing you to build up a larger stockpile. Thus, every day you can feed tree fodder means another day of stockpiling, which means one fewer day of feeding expensive hay in the winter.

It's all well and good to plan and plant ahead for summer slump and drought, but how many trees should you plant? I wish I could tell you, and of course it all depends on your conditions. While little research has been done on the amount of leaf fodder produced by pollarded trees, a rule of thumb given by a veteran New Zealand tree pollarding farmer is that a mature willow, cut every five years, will produce enough fodder when cut to feed ten cows for a day.[24] If you break that down to two cow-days per tree per year, you have a useful starting point to do your math. Using those numbers, if you have one hundred cows and want to get them through a thirty-day spell of no forages (or supplement a quarter of their feed for four months), you will need fifteen trees per head.

---

[23] Steve Gabriel, "Quantifying Nutritional Value and Best Practices for Woody Fodder Management in Ruminant Grazing Systems," Final report for SARE Project FNE19-930, 2021. https://projects.sare.org/project-reports/fne19-930/

[24] New Zealand Poplar and Willow Research Trust, "Pollarding Willows from Within the Tree," 4:18. https://www.youtube.com/watch?v=-3hd20jAYEI&list=PLNFVgb4v_GD9cGwlcLiwTPvqm9FTO9zTP

This linden tree was planted as part of a row destined to become living fence posts. In addition to holding up wires, they will provide shade, fodder, and valuable habitat.

Whatever your needs and whatever the figure you use, it's probably best to be on the safe side and plant more trees than you calculate needing, since having extra insurance is always welcome when the time comes. Also, remember that a pollarded tree doesn't take up nearly as much space or light as a standard, un-pollarded tree, so many more can readily be fit into an acre without compromising forage growth.

Rowan Reid shares the following:

> During the 1982–83 drought, Paul Dann noticed mobs of hungry sheep sheltering in the shade of healthy Crack Willows along the dry creek beds around Canberra. He fed both lambs and hoggets a sole diet of freshly cut willow leaves and twigs for a continuous period of 6 weeks starting in January 1983. The sheep ate an average of 0.94kg (~2lbs) of dry matter per day and gained an average of just under 2 kg (~4.4lbs) in 6 weeks. Paul found that a 15m (~49') tall tree produced about 200kg (~440lbs) of fresh feed, which he judged would provide a weight-gaining, drought-saving ration of 80 sheep days per tree. They recovered well and would have been available to harvest again within a few years, when the next drought might be expected.[25]

The economic breakdown here is interesting to consider. Because you can propagate these trees yourself from live stakes, the material costs are zero once you have some trees established. As an example, let's say the initial establishment costs are $5.00 per tree. We're talking trees, so let's stretch the $5.00 material cost out over twenty years, over which time period you could get forty cow-days' worth of feed. Let's say the labor required to pollard a tree is ten minutes and you assume the cost of labor is $20.00 per hour, including wages, using an ATV to get there, etc. All in all, it gives a cost per cow-day of $0.46.

This may very well be on the higher side. If you break down the numbers given by Jaime Elizondo and assume paying $20.00 an hour and needing two hours to feed 250 head of cattle, your labor costs of feed are only $0.16 per head per day. We would need to know material costs in order to give a full estimate, but given that you are direct-seeding your trees, that cost is going to be minimal.

This is really interesting when you compare leaf fodder to the cost of hay. A grassfed dairy client of mine calculated their cost of feeding heifers and dry cows (average 1,049 pounds per head) at $4.50 per head per day. That's high-quality hay,

[25] Rowan Reid, *Heartwood*, 191.

and many will have lower costs, but maintaining good condition and production made the high cost acceptable. Many producers will see significantly lower costs, in the $2.00–$2.50 range, according to the University of Missouri.[26] Either way, you can see why we want to reduce those hay costs and why tree fodder might make a significant dent in them.

Remember that feeding hay also requires labor and machinery costs. The $4.50 per head figure only accounts for the purchased cost of the hay and does not take into account the labor to feed the hay. In an article by Tim Krawiec in *OnPasture*, he gives costs for feeding hay ranging from $0.10 per head per day (bale grazing) to $0.45 per head per day (using a tractor to unroll hay).[27] Your costs will vary. When I did the math with the operator of a small dairy farm in my region, it came out closer to $1.00 per head per day because of the lower labor efficiency of feeding a small herd and having to split the cost of a tractor across a smaller number of cows. One major benefit of pollarding versus feeding hay is that pollarding requires very little investment in machinery. A small chainsaw should do, and you can leave the expensive tractor in the barn (or sell it!). It's a practice that scales down really well to homesteads and small farms and, if done systematically, can really add up to a lot of savings for large operations.

While we're on the subject of feed costs and winter stockpile, let's look at the feed cost for honey locusts to see how they compare. Earlier I used 87.5 pounds of dry matter yield per tree as an average, and I'll just round that up to 90 pounds to make the math easier. Given that the caloric content of honey locust pods is about twice that of hay per pound of dry matter, pods will act as an energy supplement in addition to providing protein, fiber, etc. The main protein and bulk needs should be met with stockpiled forages or hay, which tend to be high in protein and lower in energy. In that way, pods and stockpiled forages complement one another well.

Let's say daily intake is 10 pounds of pods and 20 pounds of stockpiled forages in a typical scenario grazing winter stockpile. The honey locust pods make up a third of the feed, and 90 pounds will last nine days, giving us three full cow-days per tree. Now, the difficulty in making a one-to-one comparison is that honey locust pods will have significantly higher energy value than forages, making ten pounds of pods more valuable than ten pounds of forages because it will yield higher animal production—especially in cold weather when energy needs are greater. So, using cow-days is a useful metric here, but by no means a perfect comparison. If anything, we're being overly conservative.

To get a (very) rough back-of-the-envelope calculation for our cost of supplying those honey locust pods, let's assume the establishment cost of each honey locust tree is $60 to give us a nice round number. It could be higher or much lower, depending on cost share, how much sweat equity the landowner puts in, etc. Often, cost share should bring this price tag far lower, as we'll explore further in the Funding section. Let's spread that cost out over twenty years of pod production, getting a cost of $3.00 per year. For reference, a grafted honey locust should bear meaningful amounts of pods by year 5. At three cow-days per year, the annual cost of feed per cow-day comes to $1.00. Again, compare this to the cost of hay, whether at $4.50 or $2.00 per cow-day, and it becomes clear that quality honey locust trees dropping high-energy pods should find a place on any farm. Plus, a honey locust tree will bear feed for generations to come while providing dappled shade to livestock beneath—a sound investment on its own which

[26] Gene Schmitz, "Calculating Winter Feed Costs for Beef Cows," University of Missouri Extension, October 20, 2022. https://extension.missouri.edu/news/calculating-winter-feed-costs-for-beef-cows.

[27] Tim Krawiec, "To Unroll or Not to Unroll: What Gives the Most Bang for the Buck When Bale Grazing," *OnPasture*, February 24, 2020.

can be made even sweeter through cost share, as we'll discuss later.

So yes, all of this can indeed add up to significant savings for the farm. These trees can allow you to reduce your dependence on costly outside feed, diesel, tractors, and tractor mechanics. You can use trees to provide feed during the times of year you need it most, which allows you to reduce your feed costs, expand your herd, or both—all while building resilience to whatever nature throws your way. That's the way to create farms that thrive for the long term, steadily producing a surplus while also being pleasurable to work and live on.

WINDBREAKS

Another benefit that trees can bring to a pasture system is serving as windbreaks. But windbreaks have not been a major focus of Trees For Graziers yet, and hence we don't have much new material to add to the body of knowledge.

The most basic decision you'll need to make when planning windbreaks is whether you want to use conifers or deciduous trees (those that drop their leaves in winter). Conifers, of course, provide a better windbreak because their many needles create a much denser screen through which

Table 4: Overview of shade and fodder species, including their best use on the farm.

	Shade	Fodder	Suggested Best Use
Honey Locust	Great for dappled shade. Moderate growth rate.	Winter stockpile.	Centerpiece of silvopasture system, providing dappled shade and winter feed.
Persimmon	Good, maintains small canopy that doesn't spread out wide, which forces livestock to spread impact between trees. Slow-growing.	Dropped fruit late summer through winter. Loved by ruminants, pigs, and poultry.	In ruminant systems, it plays a smaller, support role to honey locust.
Mulberry	Wants to grow too wide and low for optimal ruminant shade profile. Fast-growing.	A key early-season fruit for pigs and poultry. Also a high-protein leaf fodder. Best dual-purpose tree for ruminants and monogastrics.	Likely the most valuable tree for pigs and poultry. Choose clonally propagated females for fruit yield.
Oak	Good shade profile; large tree that will spread shade widely. Slow-growing.	Dropped acorns in fall. Mostly for pigs, but ruminants can use some, and turkeys and chickens can make use of smaller acorns.	Adds great wildlife value and long-term productivity. Fantastic timber species as well, if managed for that.
Apple	Wants to grow too wide and low for great shade.	Dropped fruit in late summer through late fall. Tons of cultivars to choose from.	Great for hogs; also useful for ruminants in a smaller support role to add diversity.
Willow and poplars	Very cheap and fast-growing; tall live stakes give you a head start. Willows give denser shade; poplar canopies are often a bit sparse.	Good summer leaf fodder. Willows are higher in tannins, which can be positive or negative, depending on the dosage, other forages, and livestock type.	Quick, cheap shade trees that make vigorous pollarded fodder options. Start some early so you have free propagation material. Minimal timber value.
Black Locust	Great. Fast-growing and dappled.	Small but high-protein leaves. You may not want to cut the thorny branches and leave them on the ground.	Quick shade; left as a tall shade tree that fixes nitrogen and feeds bees. Then harvested for timber after 8+ years. Backup leaf fodder source if needed. Sucker sprouts can be a concern.
Sycamore	Fast-growing yet long-lived. Fairly dense, but the tall canopy will move shade far and wide. Can be live staked.	Little value here. Might serve as a backup fodder, but not the main intention.	An inexpensive, long-lived, and beautiful tree that can serve landowners who want a simple and aesthetically pleasing option

wind does not readily pass. The benefit of using deciduous trees like willows or poplars is that they tend to grow faster, can easily be protected by tree shelters (which don't work nearly as well for conifers), and can be used for fodder as well as windbreaks. Especially in the case of deciduous trees, more rows will create a more effective windbreak. Dwarf everbearing mulberries, hazelnuts, wild plums, and other hedge-forming shrubs could be useful additions for browse and mast while providing relief from winter winds that sap energy from out-grazing herds.

One plant I do want to cover is bamboo, which is immensely intriguing as a windbreak because it does so much more than just slow down the wind. In fact, bamboo can address all of the most common issues that graziers face. The plants are a top-notch windbreak that is also tall enough to provide shade, and their leaves stay green all summer and all winter. As wild as it might seem to plant bamboo on a farm, there is a real historical precedent—river cane, which is our native bamboo species, was once the backbone of the grazing industry in the South. Professor Mart Stewart writes:

> These "tree grasses" provided the main source of winter browse for open range cattle herding. . .in the eighteenth and early nineteenth centuries. . . .These cane, which resembled their tropical relatives, the bamboo, in both size and in versatility of uses, were an attractive feed for animals and provided especially good habitat for cattle, deer, and bison as well as a host of smaller animals. . . .Cane was so essential to open-range cattle herding in the South that this important agricultural enterprise would not have thrived without it. . . .The best food for cattle in all seasons, but especially in winter, came from the ubiquitous patches and dense fields of cane. Modern studies have established that cane foliage was the highest yielding native pasture in the South. It has up to eighteen percent crude protein and is rich in minerals essential for livestock health. Settlers recognized that not only did cattle gain weight when they fed on cane, but also produced superior milk and butter; horses who fed on cane were able to work almost as well as those who were corn-fed.[28]

It's amazing how quiet and protected even small bamboo groves are.

You can see why I would be itching to start trialing projects with river cane. Unfortunately, the few small rhizomes I've gotten my hands on to date have not yet taken off, and my experience with the plant is next to nothing. It seems that the main bottleneck in utilizing cane or other bamboos is the ability to propagate it affordably enough for farmers to plant at scale.

Some folks will push back on this idea and say, "Isn't bamboo invasive?" First off, there are two kinds of bamboo—running bamboos that can spread quickly horizontally, and clumping bamboos that do not. More importantly, bamboo does not jump from place to place. If you plant it in the middle of your farm, it won't spread to another

---

[28] M. Stewart, "From King Cane to King Cotton: Razing Cane in the Old South," *Environmental History* 12, no. 1 (2007): 59-79. https://doi.org/10.1093/envhis/12.1.59.

farm a mile away. As such, it is very different from something like a Bradford pear or paulownia or honeysuckle bush. And while bamboo is tough, you can overgraze it. That's a major reason why river cane became so scarce—it got grazed time and again without adequate rest. If ever you get sick of the bamboo, you can use that to your advantage and just graze it until you've exhausted it.

One use I'd love to see explored is living bamboo barns. Because they offer such great windbreak as well as an overhead shelter, they should offer significant protection over the winter for cattle, poultry, or even pigs if they are ringed and can't destroy the rhizomes. Plant tall bamboos in the middle and shorter ones on the outside as windbreak. If you want to make a perch for poultry, just take a bamboo pole, tie it to two others, and you're all set. If cattle need supplemental feed, it's right above them. And if parasites started to build up, you can burn the "barn" down and let it regenerate from the rhizomes. And you wouldn't even need a building permit.

CHAPTER 10
Choosing Genetics

For mast trees in particular, selecting the right trees with the right genetics is very important. It's the difference between a wild honey locust with thin pods and eight-inch thorns and a selected, thornless, high-quality tree with thick, sweet pods. It also means the difference between persimmons that drop all their fruit by October and trees that hold onto their fruits through February.

If you are looking to start a dairy operation, you don't buy Texas longhorns. And if you are going to start raising chickens for eggs, you don't buy Cornish crosses. The same applies to trees, because what you choose should really depend on your goals. Thankfully, there's a huge range of genetics to choose from, and it only takes some insight and planning to do it right.

The three trees that I believe have the highest upside as livestock feed—honey locust, persimmon and mulberry—all happen to be species that generally have male and female flowers on different trees. There are some that have both male and female flowers on the same tree, but that's the exception rather than the rule. Hence, if we plant seedlings, we will get a certain percentage of males and females, though I do not know the exact ratio.

I am sometimes envious of those who work primarily with trees like chestnuts, oaks, or hickories, since all of their seedlings can produce crops. Propagation is very straightforward in those cases, seedlings are inexpensive, and seed is easy to gather in large quantities.

But that is unfortunately not the case with the main silvopasture trees. If we planted all seedlings, it would be a bit like when I consulted for a no-kill Hindu dairy. There's a reason why commercial dairies don't keep bull calves—you don't get much milk from them, and they eat a lot of food. Having an equal number of male and female cattle meant that the farm, though selling its milk for enormous prices, also had terrible productivity. Half of the animals weren't doing what most dairies expect. If we want our trees to pull their weight, we need to look at clonal propagation.

This is not a new problem for people, so there's a whole array of ways to clone a plant. The simplest way was discussed earlier—live stakes, also known as hardwood or dormant cuttings. That's probably the easiest propagation method out there, and it's great for those species that work that way. Unfortunately, that list is pretty short, and for the species we work with it's mostly confined to willows, poplars (cottonwoods and such, not tulip poplars), and sycamores. You just take a dormant piece of wood, pop it in the ground, and it starts to grow in the spring.

Softwood cuttings—like hardwood cuttings—are branches gathered during the growing season. They are trickier to handle, but will sometimes work where hardwood cuttings do not. At this point, we don't actively propagate any silvopasture species from softwood cuttings, though we've tried with several—especially mulberry.

The percentage of success was just too low, and the process was too finicky.

Root cuttings are similar, but instead of working with a branch, you're using a dormant piece of the roots. The advantage is that you already have roots to work with, but the disadvantage is that they are much more laborious to gather than pieces of branches and often need to be taken from younger trees that have actively growing roots near the trunk. If possible, it's better to take hard or softwood cuttings, but root cuttings are a good alternative for certain species like black locust.

Perhaps the most common way of commercially growing improved trees is through grafting. Grafting is the process of taking a piece of a known, high-quality tree and fusing it to a tree with less desirable characteristics—or characteristics that are good for growth but not for fruiting—called the "understock." Every single apple you see in the store came from a grafted tree. For instance, the first Granny Smith apple was a chance seedling found by Maria Ann Smith in Australia, and every Granny Smith tree currently in existence is a clone from that single tree.

One downside of grafting that's worth mentioning is that this is the only vegetative propagation technique mentioned here that does not produce a full clone. The top of the tree has different genetics than the bottom. In most cases that's completely fine, but an actively grazed pasture is a pretty challenging place to get trees established and they are considerably more likely to sustain damage from rodents, deer, cattle, or what we call heavy metal disease—which is what happens when a tractor operator tries to get a little too close with the brush hog. If the grafted portion of the tree gets chopped off and the tree comes back from the roots, you'll have whatever understock was used, not the expensive graft you paid for. That's why I would prefer to have full clones whenever possible.

This weeping cherry perfectly illustrates the makeup of grafted trees. The cultivar that weeps is not strong enough to stand on its own, so it needs to be grafted to a normal cherry understock, and the graft height determines where the weeping starts on any particular tree. In this case, the single shoot growing straight up is a bud from the understock that woke up and decided to keep growing—until I later pruned it off.

The last main method I'll mention is tissue culture. This is a more high-tech, controlled, and sterile way of propagating trees very fast—you could theoretically take one tree and make it into a million in little over a year. It's more expensive than most of the other methods and requires a specialized lab and significant skill, but if you wanted a lot of trees really fast, this is the route to take, at least for those species that can be propagated this way.

As I write this in 2025, the availability of high-quality, clonally propagated, silvopasture-specific nursery stock is probably the biggest limiting factor to the adoption of silvopasture on a larger scale. While you can and should get started with trees for which the specific genetics are either figured out (hybrid poplars, willows, chestnuts) or don't matter as much (black locust for quick shade), the benefit from waiting until better persimmons, mulberries, and honey locust are available is significant. While for several years I avoided running a nursery, having too much on my plate already, TFG started work on propagating silvopasture-specific trees in earnest in 2023. It's been a long ride with lots of learning along the way. It's slow, too, because we can only work at the pace of trees. Thankfully, we've made significant progress, and there's finally light at the end of the tunnel. It looks like the next several years should see a steady increase in the availability of high-quality, clonally propagated trees—first mulberries and persimmons, and then honey locust.

In the first edition of this book, I recommended that folks start by planting seedling trees and later come through and graft them once the trees were established. I can no longer recommend that, at least for most folks. While it can certainly work—I have a whole scion nursery that is made up of trees planted exactly that way—it is not an easy route. Grafted trees require significant work and attention to make sure the graft survives and becomes the dominant part of the tree. That is one thing to do in a nursery, where trees are lined up a foot apart and you can go down the row, steadily checking in on each tree. In a pasture situation, the trees are further apart, they are different ages and sizes, and each one is protected by tree shelters that need to be removed to cut off sprouts from the understock that would otherwise rob energy from the graft. You've got cattle and deer to contend with, as well as ladders to haul around if you're grafting above browse height. It's a real doozy of a trial by fire for those not already experienced with grafting, and even if you are experienced, the logistics might all add up to make it not worth the time.

Thankfully, there are better options on the horizon. Back when the first edition was written in 2021, we were a long way from having any significant number of grafted trees available, and the very few other sources that sold grafted honey locust or persimmon were quite expensive, at least to do at scale. Barring another nursery popping up and making these trees available at half the price, grafting in the field made more sense, at least on paper. Our nursery has come a long way since then, in fits and starts, and it's so much easier for us to care for trees in a nursery setting than in a pasture, allowing us to produce at higher scale and lower prices. That's the goal. If right now you have to pay fifty dollars for a three-foot grafted tree, our goal is to offer a six-foot tree for just twenty-five dollars when bought in larger quantities. I'm confident that day will come.

Until then, what I recommend is to hold off on the honeys, persimmons, and mulberries. Start with shade and fodder trees. They are cheap, easy to find, and very forgiving. If you need shade across your farm, start a bunch of willows and poplars and then propagate more from there. If you can't keep them alive, keep trying before you plant more expensive trees. It'll save you a lot of money and hassle later.

A question to consider is how complex you want your tree system to be. A silvopasture system can be as simple or complex as you want, with each option having strengths and weaknesses. For example, the well-documented planting at Virginia Tech is dominated by honey locust and black walnut. Both of those are great trees for silvopasture and make for a simple system that easily integrates into any grazing operation. It's also much easier to control variables for research in a simple system, which makes them very attractive for a research institute. For most farms, however, the risk is that a low diversity of trees is a greater target for pests, and losing one species through an introduced pest or disease would be a major loss.

Mark Shepard, author of *Restoration Agriculture*, is on the other side of the spectrum, having created an immensely diverse agroforestry system on his home property at New Forest Farm. That system is built for resiliency and biodiversity but is not geared toward producing the most of any one product.

My general recommendation is to start with relatively few trees and add more diversity over the years as you build out the tree system on your land. Start small, and start now. Get just a few trees in the ground, see how they do, and plant some more next year. Keep them close to the house or the barn, where you're most likely to see them and interact with them often. That's the best way to earn that valuable first-hand experience, which will pay significant dividends down the road.

CASE STUDY
Fiddle Creek Dairy

Fiddle Creek Dairy holds a special place for me because I have experimented and learned much about silvopasture implementation through my work there. Fiddle Creek is where I was first asked how a farm might establish trees in an actively grazed pasture and where I first started experimenting with means to accomplish that.

Fiddle Creek is a very small dairy farm making high-quality yogurt from grassfed cows. The issues they faced when we started discussing a silvopasture plan were the same issues many in the region face: the need to feed hay in the summer, large winter feed bills, lack of shade in the pasture, and cold winds throughout the winter. Because of their small land base, it isn't economical for them to make hay, so they are dependent on purchasing hay from others. Usually, they begin feeding hay in the summer and continue all the way until the following spring. Molasses is used to boost the energy intake of cows. A homemade portable shade shelter was used to provide some shade in the pasture, although unless it was moved daily, the spots underneath would be destroyed and create havens for weeds long after.

This picture was taken at the end of the second growing season, and we started with very small trees compared to what we use today. The faster-growing species —black locust and hybrid willow—were averaging over twelve feet tall.

Because NRCS did not—and still does not—offer cost-share for silvopasture in Pennsylvania, we had to get creative when applying for funding. I was able to locate a program that was designed for streamside buffers but was flexible enough to be used as a silvopasture, albeit a dense one, given that the program required trees to be planted at a density of 155 trees per acre, much higher than I would typically use. This creative source of funding meant that the farmer had to pay nothing out of pocket for the silvopasture project. This was especially valuable because most of what we were doing was completely experimental at the time.

We settled on planting roughly 20 percent of trees in honey locust, with another 20 percent or so in persimmon. The other trees would be fast-growing fodder trees to alleviate the need for hay and quickly provide the shade their cows so badly needed. Black locusts, hybrid poplars, and hybrid willows have grown incredibly fast there, while mulberries, persimmons, and honey locusts have grown a bit slower.

In addition to planting trees for the sake of livestock, we planted some trees as means of income diversification. A small apple orchard was planted, and a small number of grafted hickories, pecans, and chestnuts were added. Since the farmers already sell their yogurt through farmers markets, they have a ready outlet for small volumes of these tree crops.

One side benefit we are interested in observing is how the nutrient cycle is altered by the presence of trees. The record rainfall of 2018 seemed to leach many nutrients out of the soil. We expect to see improvement in forage growth and nutrient content from nitrogen fixation by black locust and the movement of nutrients from deep soil horizons to shallow soil levels through annual dropping of leaves and die-off of fine root hairs near the soil surface.

Here electric fencing is simply strung along the row of trees so that the tree row forms a paddock boundary. It's simple and effective. You can see, though, that it allows grass to grow up underneath the wire, because stock are reluctant to graze there.

We have learned quite a bit from this planting. One thing I would do differently is take a much slower approach. In hindsight, I can see that I had the same ambition I now coach many of our clients down from. You get the tree-planting bug, there's funding that will pay for it, and you want to go all-in to get shade across the farm as soon as you can. I also did not know whether there would be other sources of funding available in the future, so we planted as much acreage as we could under the grant and bit off more than we could chew, especially given the high density of planting. Had we been able to plant at, say, 30 instead of 150 trees per acre, it would have been much easier to manage.

Similarly, I made the mistake of not building in a budget for post-planting tree care when we applied for the grant. That's something our team does with every grant now so that our crew can visit each farm twice a year for the first several years, since the more preventative care you can give the trees, the better success rates you'll see. And while cattle can help somewhat by keeping vegetation down, they also put pressure on shelters by rubbing on them, accidentally running into them, etc. Things happen and need to be corrected before they turn into larger issues.

Another challenge we observed is how to protect a row of trees when the fast growers are twenty feet tall and no longer need electric fencing, but some of the slowest growers like persimmons or hickories are still only five feet tall. It would be really nice to be able to remove the fencing and open things up without leaving the small trees vulnerable. This is a major reason that we're moving to planting taller trees—especially of our slow-growing species—so they can get through that challenging first phase more quickly and uniformly.

The first year we saw significant pods on honeys was 2025. These trees were six years old from seed and had been planted as small seedlings less than twenty-four inches tall.

The willows grew like champs, even though it wasn't a wet site. These were planted as small live stakes, not the ten-foot poles we now use, and they still got to thirty feet tall in just five growing seasons.

Photo courtesy of Costa Boutsikaris.

SECTION 3
How to Establish a Tree

Now that you have determined which tree species will help you address problems or unlock new opportunities on your farm, it's time to determine how exactly to establish those trees.

Photo courtesy of Costa Boutsikaris.

CHAPTER 11
Things to Know Before You Plant

TIMING

First off, when should you plant your trees? This depends somewhat on your region and what tree stock you use, but the simplest rule of thumb is that the trees should be dormant and the ground can't be frozen. Generally, this means that fall and spring are the best times to plant.

Spring generally has a longer planting window than fall. Say you're in upstate New York or Minnesota and you're receiving trees grown in Tennessee. Those trees won't be fully dormant and ready to ship until late November, and your ground might freeze up in December. If that's the case, you have a narrow window of opportunity to get your plants in the ground. If, on the other hand, you're in South Carolina and the ground never freezes, or it only freezes for short periods in January and February, you have a nice, wide planting window. Most bare root nurseries will wait to ship to you until your region is ready to plant. Just keep in mind that far northern nurseries can send stock earlier in the fall than someone far south, but it might be late March or later until their ground thaws and they can get the spring shipment out. If you have orders from nurseries around the country, that can make it challenging to get all your stock at the same time, especially during certain windows.

Both spring and fall planting have their advantages. When planting trees in the fall, moisture stress is typically less of a concern because the trees are dormant and won't need much moisture until they break dormancy in the spring. If you apply a thick layer of mulch before spring, you will trap winter moisture and give the tree a solid start to life in its new home. The only reason I would water my trees in the fall would be if it was bone dry when planting and the soil itself would dry out the roots. Conversely, trees planted in the spring will start growing much sooner, giving you less opportunity for rainfall to help you out. We've now had several seasons of unusual spring drought that stressed trees planted in April and early May.

As previously noted, spring typically offers a longer planting season than fall. Some bare root nurseries don't ship in the fall, either, so you have more tree sources for spring planting. Folks who have a mild winter with minimal ground freezing have the best window of planting, and for folks with severe winter weather, you probably want to wait until spring to give your newly planted trees the best chance for success.

When planting in wet areas, I've found that it's best to wait until spring, since frost heave can cause trees planted in the fall to have their roots pushed up and out of the soil over the course of the winter from the repeated freezing and thawing of the soil. It's no fun to come back to the trees you worked hard to plant and see their roots exposed six inches above where you planted them. Thankfully, this is typically not an issue on upland sites.

Whenever you plant and whatever your source, you'll need to plan and order your plants well in advance. If you intend to plant on Friday, don't place your order that Monday. Nursery stocks are highest in the fall after the tree growing season is done, and many trees will sell out by midwinter. The pickings can get pretty slim if you're ordering in March or April, and many nurseries will sell out of their best stock even before fall planting season arrives. So plan ahead and place your order as far in advance as you can.

When your trees arrive, caring for them properly will give you the highest chance of success. Store them in a cool, dark place, like a barn or unheated cellar. The roots should stay moist, and the tops should stay dry. We like to pack the roots in moist sawdust, sand, or something similar.

CHALLENGES TO PLANTING TREES IN PASTURES

When establishing trees in active pastures, there are multiple forces to compete with. First, you have all the regular pressures typical of reforestation plantings, like deer browse, buck rub, rodent girdling, birds dropping weed seed inside tree tubes, etc. It's not uncommon to see a conservation tree planting that has turned into a graveyard of white tubes, with all of them either empty, leaning, or already fallen over. It's only with real care and regular maintenance that we've seen stream buffer plantings in our region generally hit survival rates above 70 or 80 percent.

On top of those regular pressures, livestock can browse, rub, or trample, and dense forage stands will compete with young trees for water and nutrients. There are also more constraints than in a conservation planting, because you still need to move cattle, mow, unroll bales, etc. Almost all of our clients have been organic, whether certified or not, which eliminates herbicides—one of the main tools used in conservation plantings. And all of this needs to be done without breaking the bank.

This is the most common method of tree establishment in pastures in my area. It works, but it will cost a pretty penny. The materials alone will run close to $100.

There are a thousand different strategies for getting trees into pastures. None is perfect, and each has its pros and cons. The most laid-back option is to sit back and let whatever trees want to grow up do so. The results will depend mostly on what species you have in your area. You could also go out and densely plant ten times the number of tree seedlings you eventually want, knowing that many will die, and then thin trees later as needed. Any number of devices and tools can be used to protect trees. People use all kinds of things as posts, including T-posts, fence posts, metal pipes, sucker rods, rebar, and conduit. I've seen trees protected by chicken wire, woven wire, barbed wire, hog panels, and even round-bale feeders. For weed control and reducing competition, you could use wood chips, nice decorative dyed mulch, spoiled hay, plastic, herbicides (conventional or organic), cardboard, or a weed eater. If you're not one who suffers from OCD, you could even pile brush and thorny plants around each of your trees to keep animals off. The limitations lie mainly in your time, creativity, and available resources.

This is a great alternative to using two continuous wires. A single hardwire fence runs the length of the tree row, and trees are individually protected by polywire squares held up by fiberglass stakes. This keeps much more grass in production and reduces the need for maintenance. You could even prop the fence up to move livestock through if necessary.

While there are countless ways to establish trees in pastures, large mortality rates have been the rule rather than the exception for most farm-based tree plantings. That is the clear picture I've seen after countless farm tours and client consultations. Assuming you're successfully able to keep livestock away from the tree, the main cause of mortality I see is what you could term Green Death. Green Death is a catchall term for what happens when competing plants overtake a young tree, thick vegetation offers such a perfect habitat for rodents that they move in en masse and kill your trees, or both. It usually happens when a large area around the tree is fenced off to keep cattle away from a tree and the grass inside grows up unmanaged. Yes, everyone starts out with fine intentions of returning to mow or weed whack the area regularly, but come July, nobody wants to spend sweaty hours in the sun messing with fencing and trying to take out overly mature grasses while avoiding the tree (if you can even find the tree anymore). So most trees get very little aftercare, then die, and subsequent tree planting dreams get squashed because of bad experiences. I'll bet you that's the story on nine farms out of ten. But it does not need to be the case.

While they may look cute, voles and other rodents kill more trees than livestock do.

TREE SHELTERS

What I will lay out here is the method that has worked best for Trees For Graziers and our context after lots of trial and error and has proven to be very simple, reliable, and cost-effective. It takes almost zero space out of production (only the size of a dinner plate) and requires comparatively little aftercare. It is very well suited to getting exactly the right trees established exactly where you want them and quickly getting them above browse height. Most importantly, it allows you to establish trees in an actively grazed pasture, rather than having to take the pasture out of production for a number of years while the trees are getting established. While adding trees in an active pasture requires more protection of the trees (and hence expense), removing pastures from grazing for five or more years is simply not an option for most producers.

When it comes to getting a tree established in an active pasture, I have found no tool to be more useful and versatile than the Plantra tree shelter. I stumbled upon them several years ago in my work planting stream buffers. Initially, the main draw for me was that the stakes were much lighter than wooden stakes and hence easier to carry to remote and tough-to-access riparian zones. Plus, the fiberglass stakes wouldn't rot and crack like the wooden stakes tended to.

Now that we've tested the Plantra shelters in a silvopasture context as well, a couple other benefits have become clear. The flexibility inherent in the fiberglass stake allows the shelter to withstand moderate rubbing by cattle, much more than a wooden stake would. If you're establishing trees in a sheep pasture, that might be all the protection you need, though an additional metal stake would be welcome to protect against rams. Cattle, however, tend to be even rougher on the shelters, which is why shelters in cattle pastures need protection for several years while the trees get established. And here is why I love the Plantra shelter so much—both the plastic shelter and the fiberglass stakes are electrical insulators. That gives you endless options and flexibility in protecting the shelters from rubbing and gives graziers the freedom to get trees started with minimal changes to the grazing system.

Table 5: Tree protection measures

	Pros	Cons	Notes
No shelter, protected by electric fencing	Cheap. Suitable for dense plantings where protecting each tree is cost-prohibitive. Low-risk option since inputs are low.	Will require taking a larger area out of production. Much more difficult to do in the middle of an actively grazed pasture. Tougher to control competing vegetation around each tree, leading to slower growth and higher mortality rates.	A good route for high-density plantings such as windbreaks, hedges, and timber stands because of low per-tree cost.
Plastic tubes	Provide greater protection to trees; easiest way to establish trees in an actively grazed pasture. Encourages formation of a single stem, which tends to be useful in silvopasture. Easy to connect electric fencing to.	Cost is often more than the tree it protects. Tubes can create shelter for voles. High humidity inside tubes can cause a problem for sensitive species like chestnut or apple. Requires removal and recycling/disposal. Trees get leggy inside and need support for several years after they emerge. Rubbing at the top of the tube can cause damage.	My preferred method for low-density plantings. Probably anything under 80 trees per acre.
Mesh wraps	Inexpensive, very reusable. Well-ventilated. Wraps close to the tree, so there's no space for rodents to get in between wrap and tree, which can be a problem in tubes.	Doesn't offer space for leaves, so it's exclusively something for wrapping around trunks. Would not work with an overpass system.	Intriguing option for trees that start above browse height when combined with electric fencing. As we plant more trees 6' and taller, this will likely get heavy use. When cut into short sections, makes great vole guards.
Wire cages	Readily available on most farms. Infinitely reusable. Trees can spread out more naturally than in a tree tube, though only if the cage is large enough that the tree does not grow through. No electricity needed	Trees tend to grow out of cages, unless the cages are set several feet from the tree. Much more labor-intensive to provide aftercare, since you need to remove the cage each time. Trees tend to be slower-growing in my experience. Creates a larger zone taken out of production compared to tubes. You'll need to keep vegetation down within this zone, adding to the labor.	I think this can be a good option on very low-density plantings, probably under 20 trees/acre, especially if the farm already has materials laying around. I would exclude a large area, probably 3 ft from the tree, so that the tree can expand nicely and not grow into the cage. Importantly, you need to keep competition down, so I would only do this if I could also use a heavy layer of mulch around the tree, probably 2 ft in diameter.
Wooden stakes	Inexpensive. Will naturally decompose and don't require removal.	Tend to bend or crack when rubbed against, requiring lots of repairs. Also, some will almost certainly rot before the tree is ready to stand up on its own Don't provide electrical insulation.	Can certainly be used, but don't lend themselves to silvopasture as well as fiberglass stakes.
Fiberglass stakes	Electrical insulator. Flexible, can take some rubbing. Can be reused several times.	More expensive than wood. Require removal to prevent trees from growing around them.	My go-to for silvopasture. The window of time between when the tree is strong enough to stand on its own and when the tree has swallowed up the base of the stake is pretty narrow.
T-posts	Readily available on many farms. Infinitely reusable. Stronger than wood or fiberglass.	Not insulators, though there are insulator clips made for T-posts. Heavy to transport and install. Higher up-front cost. Require removal to prevent tree from growing around them.	Likely best used to anchor cages. Could be used to support plastic tubes wrapped with barbed wire.

Table 6: Types of trees for planting

	Pros	Cons	Notes
Bare root trees	The easiest way to get quality genetics anywhere, because they're easily shipped. Much easier than potted trees for those who are certified organic.	Require more care in handling than potted trees. Have a shorter window for planting, especially in northern climates. Trees over 5–6 ft get much more expensive to mail.	The go-to option in most cases. Be sure to store trees well and protect roots from drying out.
Potted trees	Have a much longer planting window, especially in the fall. Are less fragile because their roots are covered, meaning less risk of drying out.	More expensive in general, and much more expensive to ship via mail. The soil isn't allowed on certified organic farms, unless the potted tree is certified, which is really hard to find.	Requires you have a source of quality trees nearby, or ship in large quantities to bring the cost down. Very useful for contractors, as it allows a much longer planting season. The fact that it's forgiving of some neglect makes this interesting for DIY applications.
Cuttings	Very cheap Can propagate your own	Only works for a few species, like willows, hybrid poplars and sycamores. They typically don't start growing quite as fast as when they start with roots.	A great option for the species it works for. If you want to use these species at any quantity, get some parent trees started early for future propagation.
Direct Seeding	Cheapest option. Allows trees to establish roots undisturbed.	Works best for larger seed. Much trickier than using seedlings.	Worth experimenting with for trees like walnuts, hickory, oaks and chestnuts. You'll need to protect the seeds from rodents.

CHAPTER 12
Tree Planting and Establishment

What follows is our recommended step-by-step method of establishing trees with the Plantra tree shelter:

1. Lay out where trees will go, using stakes or flags
2. Once you are set on the location of the trees, pound in stakes
3. Dig holes
4. Plant all trees
5. Install trunk guards
6. Lay out and install tubes
7. Install mulch
8. Protect shelter as needed

The first step is to lay out where you want each tree. A polywire reel is shown here as a backup option for setting straight lines (you can even mark distances on it if you're planting trees at regular intervals), but we tend to use 200-foot tape measures.

Before I pound in my stakes, I do a visual check to make sure all stakes are where I want them. To do that, put the stakes in the ground just far enough that they stay upright and double-check row spacing and straightness.

Drive the stakes down to the line indicated.

In my experience, a simple shovel works better than an auger for planting small trees. I tend to prefer a drain spade, but a standard shovel will do the trick and is better for trees with large, wide root systems.

With Plantra shelters, there will be a stake pounder included to install stakes.

You want the hole to be roughly 50 percent deeper than the roots of the tree. What you absolutely don't want is for the tree roots to bend up in the hole. If your soil is rocky and doesn't allow a nice deep hole, it's better to trim the roots a bit than stuff them all in. Notice that the hole is dug to within an inch of the stake. You want your tree to be positioned roughly one to two inches from the stake so the tree tube can be centered over it.

Install your tree, which in this case is a honey locust. Be sure to keep it close enough to the stake. If it is more than about two inches from the stake and you have to stretch the tube to cover the tree, the tube will tend to rub the bark off the tree.

Because of the damage that voles and other rodents can cause, we like to protect the base of the tree itself with a vole guard. Pictured here is a spiral wrap, although there are other options as well. We also put a mixture of blood meal and a granular castor oil product inside each tube going into the fall as an added insurance measure to discourage burrowing rodents.

Close-up of spiral wrap.

Figure 51 is a close-up of the vole guard options we've used. They are usually sold in lengths of two to three feet and can be cut down to nine inches or so to protect seedlings. We now use the black mesh mostly because it has more ventilation, takes up more space in the tube (which should discourage rodents), and is easier for the tree to push out as it grows.

We started by using the left guard—a vinyl spiraling wrap—and have since moved to the right guard due to the increased ventilation, soft edges, and ability for trees to easily push it open. The middle is a commercially available guard, but it's tighter and more restrictive of airflow than I prefer.

Install the tube. We generally only install the bottom three ties when we're protecting the tree with electric fencing. That allows the tube to be slid up and down much quicker for maintenance than if you install the top tie, because it invariably slips off and needs ten seconds to reinstall, which adds up when you have a thousand trees to check up on.

Install mulch. This picture shows two five-gallon buckets' worth of wood chips, which is our standard. You don't need to go that heavy, but the tree will benefit from lots of mulch. The more the merrier. Because the tube keeps the mulch from direct contact with the tree bark, you can pile it high without worrying about smothering the bark.

Protect the tree shelter from rubbing. This shows just one way to do it—wrapping polywire around so all sides are protected. You can tell it was a wet and muddy day when I took the pictures. Not fun to work in, but trees tend to love it.

What we love about this sequence is that by first establishing the tree shelters, you can:

- Determine exactly where you want each tree before planting, so you can avoid the labor of moving trees later if the placement was off.

- Do some of the work before your trees ever arrive, so that when trees arrive all you need to do is stick them in your pre-dug holes, slide the tube over the top, and mulch. This also allows you to do the bulk of the work in a slower time that fits your schedule, rather than racing to get everything done in a short time period after the trees come.

- Give the shelters a dry run, so you can determine whether you can protect the shelters from your livestock before you even plant the trees. This is especially helpful when you're first getting started.

We have made several changes to this process for our own plantings, which you can choose to adopt as well. As we moved from planting smaller seedlings to larger and larger trees, now regularly six feet tall with large root systems, it made sense to make the switch to drilling holes with an auger. We made the leap to buy a mini skid loader with an auger on the front, which is the fastest means of augering holes and allows you to see exactly where the auger is, which is tougher on a tractor with a rear-mounted auger. But because we're now augering instead of hand digging, we've swapped steps so that we dig first, then pound the stake in (ideally in solid soil right next to the hole), or we keep the auger shallow enough that there's still undisturbed soil at the bottom of the hole for the stake to bite into.

Another change we've made to our process is to use aluminum wire instead of polywire to wrap around tree tubes. We started with polywire because that's what we had on hand, and it worked, but it tended to arc and short more when we were tying one line to another. Aluminum wire gives a better connection and holds its form better.

Now, tree shelters alone are not enough to protect your trees from cattle, and additional protection is necessary. Because protection is very tightly linked to layout, those specifics will be covered in the next section.

ESTABLISHING DENSE TREE ROWS

Establishing a dense row of trees—whether for windbreak, browse, or maybe a dense timber planting—is different from establishing widely spaced trees. For one, you'll want to reduce or eliminate weed competition throughout the whole row, rather than in individual spots around each separate tree. Second, it doesn't make sense to protect each tree with a tree shelter, because there are so many trees that the cost would be prohibitive. Instead, all the trees are protected as a row, usually with electric fencing.

What I can't stress enough here is to do a thorough job of site prep before planting rows of trees. Even though you may be in a hurry and rushing to get the trees in, it'll save you a lot of time in the long run to do the prep work well. You can use mowing, scalping, plowing, discing, herbicide (organic or conventional), pig snouts, or any combination you choose—just make sure it gets done. I can speak from experience that when we skimped on this step, survival rates and speed were a fraction of what we wanted.

The most interesting way to prep for a row of densely planted trees might be to mow and windrow a thick layer of grass where you'll do your planting—just like you would when haying, but don't bale it. Just leave it in place to smother out the competition, mulch the soil, and trap moisture in preparation for planting. Then, when planting day comes, you can use a big coulter on a tree transplanter (picture a beefy vegetable transplanter pulled behind a tractor) to slice through the debris, plant the seedlings, and pack them in, with the mulch already in place and moisture ready to nourish your little trees.

While a transplanter works very well for high-density plantings, we haven't found them to be useful for the low-density, savanna-style plantings that are our bread and butter. I wouldn't want to plant a fodder block by hand without a transplanter, but big trees spaced twenty feet or more apart and lined up in a grid is a very different story, and we find that digging individual holes makes more sense and allows us to give better care in that context.

The method I present certainly isn't the only one. Way back in 1929 the book Tree Crops showed this picture with a subtext that read as follows: "Tall-headed pecan trees planted by owner in cow pasture of rented dairy farm. The two stakes support the tree. The barbed wires keep the cows from rubbing the stakes. The pieces of old rubber hose by the man's finger protect the tree from the stakes. This invention is freely given to the public. The trees were mulched and manured. They are thriving in the pasture of a rented farm. No overhead cost. The latest improvement in this technique is to dig a two-bushel hole above the tree and plow furrows leading water to it for shower irrigation. Try it." (Photo by J. Russell Smith.)

OOPS! WHAT CAN GO WRONG?

It's good to learn from your mistakes.
It's better to learn from other people's mistakes.
—Warren Buffett

People often ask me why so few people plant trees in their pastures. There are a bunch of reasons, including lack of funding, lack of education, lack of patience, and a culture that wants quick fixes. But one key reason is that it hasn't been easy—mortality rates have tended to be very high. My goal is to make it simple and almost foolproof, but it's taken a lot of trial and error to get to where we are, and it's still not foolproof. Some methods we tested have worked well, and I share those in this guide. Others have been complete flops. Some work well in one setting, but not in another.

I believe that repetition and constant feedback are what allows us to learn so much and make steady improvements; they are likely our greatest gift to the field of silvopasture. To my knowledge, there has never been a silvopasture contracting company that has consistently planted and then regularly followed up on dozens and dozens of farms. When you're only planting on your own farm, you tend to make do with what you have and squeeze the tree planting in between the countless other chores and projects a farm demands. But by doing this as a service, we've had to up our game significantly. Our plantings need to succeed at a high clip, every time, or we don't get new projects—especially since all of our projects happen in a small area and most of our new clients know several folks we've already planted for. So the stakes are high for us. Not only that, but by returning regularly to every planting to check in on every tree, we see what has worked, what has not, and what has worked on one farm but failed on another. And because we do everything from writing the plan to growing the trees, planting and protecting them, and then doing the maintenance, if there is a problem in any part of the chain, we have the agency to fix it.

In service of allowing you to learn from my costly mistakes, so you don't have to make your own, what follows is a list of "oops" moments:

- Tried skipping installing electric or barbed wire on shelters to see whether they could hold up to the cows without protection. We had a few broken stakes, but the main issue was that as the cows rubbed against the tree shelters, the stakes moved back and forth, cre-

Table 7: Options for reducing competition between trees and forages/weeds

	Pros	Cons	Notes
Wood chips	Great weed suppression when applied heavily. Creates fungal environment that trees thrive in. Often cheap or free.	Bulky and labor-intensive to distribute. Requires machinery access. Seems to create some vole habitat, but not terribly. Can wash away in a floodplain.	This is our standard method. As with other mulches, apply a liberal helping to each tree to provide long-term weed suppression and hold moisture.
Grass clippings	Available for free. Requires no transportation of material. Works great for long rows of trees, where strips can be mowed and grass raked. Supplies fertility to the trees.	Not available in the winter months when planting often takes place, so you need to think ahead. More likely to harbor voles; a long row would provide lots of cover.	Mowed grass can be raked up against trees. Doing this in the spring is good timing because there is plenty of grass and the heat and drought of summer is still in the future. The longer the grass, the easier it will cut and rake. Useful option if wood chips are not available or logistically unfeasible.
Manure	Available for free on many farms. Will fertilize the trees as well as providing weed control and moisture. Probably deters voles.	Bulky and labor-intensive to distribute. Requires machinery access. Will contain seeds.	A great option for any farm that needs to spread manure anyways. Likely best when mixed with bedding. I wouldn't apply fresh manure in the first year unless mixed with plenty of bedding.
(Spoiled) Hay	Available for free on many farms. Will fertilize trees. Chunks are fast and easy to apply.	Creates vole habitat, which will need counteracted. Will contain seeds, which may sprout from the hay and reduce the effect of suppressing competing vegetation. Bulky and heavy.	A good option on many farms. Just make sure to provide solid vole protection if you use hay.
Coconut mats	Requires very little labor and no machinery to apply. Very simple logistically. Good weed suppression from large mats.	Creates vole habitat, which needs to be counteracted. Needs to be bought in and can be tough to source. Livestock will mess with them if they are bored and mats aren't hidden by forages. Add no fertility.	Can be useful in difficult-to-access areas where wood chips are difficult to apply. Could be useful from a contractor/project management perspective, because you know exactly how many to buy/transport to a job site. We've stopped using them because of the vole issues.
Conventional herbicides	Low-cost way to get rid of competition. Best option for deterring rodents, because it creates an area with no cover.	Kills forages and invites less desirable plants to recolonize spots. Doesn't help moisture retention of the soil like mulch does; doesn't add fertility. Requires care to not kill the tree. Tree shelters help with this.	This seems to be the standard option recommended by many in extension or research facilities, but I prefer mulch in most instances.
Organic herbicides	Can create an area with no cover for rodents. Sets competing vegetation back and doesn't tend to kill it (this can be a positive and negative).	Requires multiple applications per growing season. Organic herbicides are less effective and long-lasting than conventional, systemic herbicides. Does not help with moisture retention, nor does it add fertility.	I would prefer mulches in most cases, but organic herbicides are certainly a tool in the toolbelt.

ating a divot in the soil. Since the seedlings had just been planted and only had small roots, the divots created gaps where there was no root-to-soil contact. That was going to lead to too much mortality, so we strung up polywire after all.

- Tried using small tree stock to save on cost and labor. Don't do this. Trees have it hard enough getting established in an active pasture. Spend the extra buck or two and get something larger. I generally like to go with two- to three-foot trees as a minimum, and we are going larger whenever we can. The most expensive tree is the one you have to replace later on.

- Planted bare root trees in a floodplain in fall. This is where we encountered frost heave. Because of funding, we had to finish the project in the fall, and we didn't have potted trees as an option. To be honest, I wasn't even expecting frost heave since the soil was moderately well drained and certainly not saturated. But there was a huge difference between the trees planted in the floodplain and those planted further up the hill. A heavy dose of wood chips around the trees to insulate them from the freeze-thaw cycle might have helped, although that is a gamble in a floodplain, where a thick dose of chips might wash away.

- Used wood chips as mulch on a farm with free-range chickens. The chickens really appreciated the new scratching opportunity, and you quickly couldn't tell that wood chips had ever been applied.

- Used coconut mats as mulch. Coconut mats are really easy to install, especially for a contractor that needs to move materials from one site to another. It's simple to send two hundred mats for two hundred trees to one site, and you could move them with a compact sedan. No trucks, trailers, wagons, or loaders to coordinate. Unfortunately, voles love them because they provide both a solid roof over their head and material for a nest. They also don't break down and feed the soil like wood chips do. After seeing high mortality rates on several projects with coconut mats, we stopped using them altogether.

- Learned that sheep like to chew on rubber twist ties. Goats will almost surely do the same. Try tucking the ties away if possible. If using electric wire, consider using a "drop wire" to convey current to the ties the sheep want to chew on.

- A farmer was having trouble keeping the fencing on shelters hot, and when a couple heifers found out, they went to town on the trees. You absolutely have to keep the fence hot if you're relying on it to protect the trees, or else have animals that really respect wires.

- Learned that one weakness of the barbed wire method is that cows can get their tails caught in the barbs, and the whole tube can come off when they yank their tail free. It's usually not a huge deal, but it's no fun to have the occasional tree damaged this way or to have to reinstall shelters regularly. Consider installing four ties per shelter for some extra grip, as opposed to the usual three ties (you'll need to ask for extra ties). Or wrap each tie several times around the stake.

- Tried using an open mesh "tube" for apple trees instead of the standard Plantra tube to keep the humidity inside the tube down. The apple leaves, which are apparently really tasty, grew out of the mesh and were too much temptation for the cows to refuse. Once they got a taste for the leaves, they proceeded to yank off any tube they could. What I'd do instead is use the standard Plantra tubes but drill or cut some larger holes in them to increase airflow. We've also noted that voles seem to girdle apple trees more than any other species we've dealt with.

- Planted some small, grafted persimmons with unique genetics. Unfortunately, we had over 50 percent graft failure after the first year. Not a cheap problem for thirty-dollar trees! The trees died down to the roots, and although they sprouted, we had lost the graft, so the resulting trees were seedlings instead of the grafted trees we wanted. Persimmons frequently abort the top and shunt energy back to the roots when stressed, which is a major reason that we've pursued full-clonal propagation methods with them over grafting.

- We initially did very little to protect against voles. Unless you want to replant a ton of trees, do not skimp on rodent protection. Use at least one means of vole deterrent, but ideally use two, like a physical barrier and a smell-based deterrent like castor oil.

- Tried to save some money by using five-foot tubes instead of six-foot and lifting them up the stake a foot as the tree got taller and grew out. It could have worked, except that the tubes were also holding up the polywire and it got to be really funky with some tubes up and some down—and then they didn't all stay up when they were supposed to. Just pay a little extra, get the six-foot option, and keep the trees better protected from cattle.

- Similarly, don't try to use a stake that's too short for the tube. It's fine until the tree gets tall, but I've seen plenty of instances when a young, gangly tree grew up and, without the support of a stake, flopped over, killing the tree.

This is what happens when the stake is too short for the tube.

- Planted hybrid poplar live stakes, about three to five feet tall, in a poultry pasture with no mulch. The forages overtook them because of the high fertility, and guardian dogs chewed on whatever survived.

- We had a client give a large part of their pasture a complete year's rest, with the trees only being two or three years old at the time. The forages grew like crazy to over four feet tall, then fell over, forming some of the best vole habitat in the world. One half of the silvopasture had a rest, the other half did not, and the half that was rested had several times more vole damage than the other half. Thankfully, it was not catastrophic, because we had taken several measures to protect from voles from the beginning, but it was a significant difference. I'm all about long rest periods on forages, but know that the more cover there is, the more it will attract rodents. Consider avoiding really long rests for young silvopasture systems where trees will be most susceptible to damage.

I write this not to discourage you but as a caution. Getting small trees established in actively grazed pastures, where there are countless things trying to eat or outcompete them, requires doing the little things right if you want a high success rate. I'd rather scare you into caution than get you all excited to do something big that doesn't turn out well.

Thankfully, this is not rocket science—which is a good thing, because I would never have figured these things out otherwise. Make a good plan, follow the steps closely, watch carefully when your livestock interact with the trees, and adjust accordingly. Again, there is wisdom in starting small, gaining hands-on experience with your unique context (tree species, layout, animals, etc.) and adding trees in a series of phases over the course of multiple years. It can certainly be done, and the value these trees can add to your farm is more than worth it. You just want to make sure you get the details right.

CHAPTER 13
Layout and Protection

LAYOUT

A critical factor in tree protection long-term farm management is determining how you want to lay out your trees and what makes the most sense for your operation. If you're only planting two trees per acre, it makes much more sense to protect them individually, whereas you might decide to protect twenty trees per acre with tree tubes and electric and two thousand browse trees per acre without shelters at all, just electric fencing.

Whatever you do, I highly recommend not planting trees at random. That might be what nature does, but nature also uses thousands and thousands of seeds to grow one tree. If you're trying to run an effective, profitable farm, use rows. I once saw a project that was otherwise done very nicely, but the trees were all at random spacing. It looked good, but made it very challenging to protect the trees, since electric fencing was not an option. Because they couldn't integrate cattle, they had to mow until the trees were mature—and it's a real winding headache to mow when there's no pattern to follow.

While there are definitely wrong layouts, there's not one perfect layout or tree spacing, just as there is not one perfect paddock layout, magic frequency of moves, or universal farm design. Whatever you choose, it needs to work for your farm. Here are some basic principles to keep in mind:

- The more even the trees, the more even the shade, and hence the more even the animal impact.
- The wider the space between rows or clusters of trees, the more livestock will congregate in those areas.
- The more trees, the more ability for livestock to spread out, and the sooner trees will provide enough shade for your herd after being planted.
- The more trees, the more additional benefits you can integrate, like windbreaks, leafy fodder, winter feed, and more.
- Fewer trees mean less establishment cost, but a longer time to reach ideal shade levels, less feed, and more concentration around those trees.
- If you move paddocks frequently, you'll want shade in all paddocks. Better yet, you'll want to provide many shade options for livestock in each paddock.
- If trees are planted close together, consider a north-south orientation of trees, which distributes shade better throughout the course of the day. The more evenly trees are spread, like in a grid, the less a north-south orientation matters.

- Orientation of the rows against the prevailing winter wind can help by providing some windbreak, even if all the trees are deciduous.

- Straight or contoured rows are much easier to navigate than random spacing.

- If you are blessed to have naturally occurring regeneration of desirable species, and you are okay with their random spacing, start by protecting them. That will tend to be cheaper, faster, and require less labor than planting new seedlings. It'll also let you experiment with ways to protect trees from your livestock.

- The easiest place to protect trees is along existing fence lines. That's a great place to start, and on farms where paddocks are small, this may be all you need to do.

- However they are laid out, rows need to accommodate your farm access needs like clipping, renovation, haying, etc. You don't want the trees to create headaches in your farm management. This should be a top priority.

- A good rule of thumb is to create rows of trees that are a multiple of your largest equipment. If you want to bring a 30-foot mower through to clip your pastures, make your rows at least 35 feet wide (leaving some buffer room so as not to hit the trees)—or 60, 90 or 120 feet.

- Give yourself room for turning equipment. Make turning areas at the end of rows several feet wider than you need, then give yourself a little extra buffer beyond that. It's better to plant an extra tree later than mow over a tree you shouldn't have planted or get your equipment stuck.

- If planning to run hogs or poultry, having access to dropped feed is of utmost importance. Designing with blocks of certain crops, like mulberries and persimmons, means you can concentrate livestock or poultry access to certain areas when those crops are dropping.

Wide rows are relatively easy to establish and are often a great starting point for plantings to get some shade in many paddocks. The downside is that livestock will tend to congregate on the edges and concentrate manure and impact in those areas.

This illustration shows both the pros and cons of a denser planting. You get shade quick and can spread your impact out nicely, but at a certain point you have more shade than you need and will need to thin some trees.

While we've planted a whole variety of densities, twenty to forty trees per acre seems to make the most sense for most common applications.

One way to ease yourself into silvopasture is to start with wide rows, then fill in later over multiple planting phases. For example, initially rows can be 120 feet apart, then 60 feet, then potentially 30 feet as you see the value of bringing more trees into the pastures. Planting in phases like that allows you to reduce costs and potentially makes for easier management. Successive rows can be established once trees from the first phase are mature enough to not require protection, allowing you to reuse some tree protection.

We tend to use linear plantings, but there's no reason not to plant in clusters. Clusters can actually be easier to protect in some cases and create a nice effect of shade islands dotted through the landscape. They can also be a useful way to intentionally congregate animals in one area, whether to pattern deer during hunting season or to move livestock into an area with a lot of trees dropping at once. Or, if you're concerned about runners coming up from black locusts, you could establish locust groves. Then, rather than having them spread across the farm, the downside is contained, and within the space of that grove the trees are free to develop new shoots, giving you a recurring source of posts.

Many folks familiar with permaculture bring up contour planting, keylines, or swales. Though there are contexts where those are useful, we have not used them in our projects. We operate in a region where terrain is not steep, droughts tend to be rare, and rain is scattered throughout the year. Wood chips are our best way of watering, and after the first year or two the trees are ready to go. We also tend to operate on smaller farms where the other infrastructure (buildings, roads, fences) is already established, which makes the new establishment of larger features that require earth moving—like swales—more challenging. For this context, where moisture is not a key limiting factor in tree establishment, I think ease of access is more important than optimizing water flow to the trees.

SHELTER PROTECTION

Once you've determined where you want to establish your trees, you'll need to decide how you'll protect your shelters while the trees are young, if you are protecting individual trees with tree shelters. The best option we have found is one of several electric fencing configurations. Each method has pros and cons, and certain layouts lend themselves to one or the other.

Single Strand. The simplest arrangement is to run a single strand of electric wire along a row of shelters that also serve as your paddock boundary. The beauty of the Plantra shelters is that the fiberglass stakes and rubber twist ties are both insulators. This means that the tree shelters actually replace the step-in posts you'd otherwise have to use, potentially saving you a good bit of time on your moves. Simply attach the wires to one of the beefy rubber twist ties included with the shelters, and you're ready to move your stock. If you have trouble with cattle browsing at the top of the tube, you can string another strand at the top of the tube to deter them.

Barbed wire. It gets more challenging if you want to protect your trees with electric fencing but also want livestock to move through rows of trees. We initially used barbed wire for that, wrapping each tree shelter with about six feet of wire. It worked fairly well on one farm but failed on other operations. The farm that has seen success mostly grazes their mature cows in this silvopasture and has fewer young, rambunctious animals who want to mess with trees. They also move cows twice a day, so they rarely get bored. I would love to say that barbed wire works well enough to recommend it, but I've seen too many cases where it doesn't do the trick.

Overpass with drop wires. This method allows cattle to move freely underneath the wire and between the rows, allowing the grazier to set up any size of paddocks in any direction. It's been a real game-changer in making management easier and has become our most common method. A single strand of polywire is strung at six feet (the highest part of the shelter), allowing livestock to pass underneath. To protect the trees from rubbing, we then connect a piece of wire, typically aluminum, to the polywire and wrap it around the tube.

Underpass. If you need to get not only cattle, but also machinery, through a row of trees, you can run an insulated wire under the ground. We did this for a large dairy in Pennsylvania that needed regular access with a large tractor for clipping and unrolling hay bales.

The top of the tube has a wire tie inserted, through which we run the polywire. Then the aluminum wire gets wound around the poly and wrapped around the tube so that livestock don't dare rub.

These cattle are easily able to move between rows of trees because of the overhead wire. Each shelter is wrapped with aluminum wire.

When protecting a row of trees, remember that you don't necessarily have to use a single unbroken row. You can leave gaps in the line for cattle to pass through and connect your fencing to a hot fence on either end. Alternatively, if there are no hot wires nearby to connect to, you can use a solar charger.

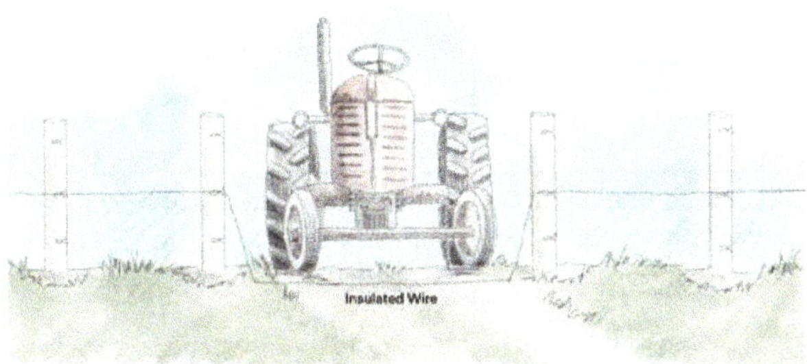

This is just one way you can get creative with wires. You could also run an overhead wire as an alternative.

If your herd is well trained to electric fencing, you may get away with using some cold wires as a deterrent, at least for short stints. However, I've also seen trees damaged badly when cows learned that the electric fence wasn't staying hot and decided to play around. So that trick should only be used sparingly.

You can, of course, also opt to not set up these semi-permanent wires and instead use polywire that you set up and take down regularly as you do for the rest of the farm, though the practicality of that really depends on your layout.

There are cases where protecting the whole row with electric fencing is impractical or you don't have access to a source of electricity. Barbed wire is an option to consider in such cases. I highly recommend starting on a small scale to see how your livestock interact with those shelters. I would also recommend reinforcing the shelter system with a second, larger stake—a T-post, wooden post, or thick fiberglass rod. Or you can use a cage, as we discussed earlier.

Taller cows tend to be tall enough to browse the top of the tubes. This tends to be less of an issue when the overpass electric fencing system is used, because the electric acts as a real deterrent. Some sites and species tend to have worse luck here than others. Adding a second line at the top of the tube can resolve this issue.

Table 8: Summary of different tree protection fencing options

	Pros	Cons	Notes
Electric, overpass	Gives the most flexibility in management. Livestock tend to graze well underneath the wires, so there's less to mow. The high wire seems to reduce the likelihood of taller cows browsing on trees as they emerge from tube.	More labor-intensive to set up. Since the tubes and stakes are required for holding the wire up above the livestock, they need to stay in place longer than the single strand system.	This has become our most common method because it is flexible and provides reliable protection from livestock.
Electric, single strand	Cheap and simple. Allows materials to be reused faster, because the tube is not necessary for holding wire up.	Livestock don't tend to like grazing under the wire, so forages can grow up thick. Limits paddock sizes and shapes, unless you reel up the wire each time. Taller cattle can sometimes reach trees emerging from tubes.	This works particularly well if you want the tree row to function as a paddock division. In those cases, it's a 2-for-1 deal.
Barbed wire	Gives flexibility if electric fencing is not feasible.	Not fun to install or get rid of. Doesn't deter livestock as well as electric. Tails can get caught in wires. Livestock seem to like rubbing gently on the barbs.	Use glasses and gloves if you choose this method. I highly recommend you add additional, heavy tree stakes for more support.

ALLEY CROPPING

Don't forget that pastures aren't the only places trees can be planted. Adding trees into crop or hay fields can be very useful for those looking to eventually convert that land to pasture. Stockpile trees like honey locusts can also be used as a means of integrating livestock into crop grounds over the winter. If you're grazing cover crops off a field in winter, it would be great to have honey locust pods there as well. As long as you space trees widely enough, there should not be a significant yield decrease in your annual crops, and they will add real diversity—both above and belowground—to otherwise one-dimensional fields. Establishing trees in areas without active grazing means you won't have to protect the shelters from cattle. Plus, cost-share for alley cropping can be significantly more robust than for silvopasture, depending on your state.

There are also scenarios where it might actually be cheaper to take the land out of grazing for several years while the trees get established, rather than provide this much individual protection. We established these protection methods in a context where land regularly leases for three hundred dollars or more per acre. If you're in a spot where land costs fifty dollars an acre and protecting them from cattle would cost an additional five hundred dollars, it might be worth haying it for several years instead. Or, if you already have more land than you need for your cattle, you can sideline those pastures for a bit while the trees get up to size. That said, in most areas deer are still a concern, so some protection is almost always needed, and there is not much cost difference between protecting from deer and protecting from cattle.

Pigs, with their rooting habits, can be especially hard on trees. Thankfully, they tend to be run on a smaller section of the farm and don't benefit much from the trees until they are providing good shade or dropping mast anyway, so it makes sense to keep them away from the trees until the trees are well established (and even then, keep a close eye on their interaction). While you can certainly find ways to protect trees from pigs in active pig silvopastures, it's just easier if they aren't a factor.

This setup has three fiberglass sucker rods around the base of the tree, with a hot wire dropped from the top of the shelter and wrapped around the rods. Photo courtesy of Josh Payne.

PLANTING PATTERNS AND SIZING

Once you have determined the species you'll be planting, you'll want to determine a planting pattern. The simplest pattern is to have only one species in each row. If you have multiple species, you can determine whether to place them at random or use a set pattern. We tend to use multiple species per row in a predetermined pattern. We'll often alternate fast-growing and slower-growing trees, so we know that every other tree will get taken out or pollarded as the trees mature.

This tree was planted just a month or two before the picture was snapped. What a difference it makes when we can get trees above the tube from the very beginning. Photo courtesy of Bryan Hoffman.

As we have the ability to work with larger and larger trees, I believe that we'll end up planting at lower densities and likely using a larger percentage of our long-term species like honey locust, persimmons, or oaks and fewer of our quick shade species like willows or poplars. If we can plant a seven-foot-tall honey locust that will start giving shade in three years, why plant a black locust that will provide shade in two years but will get removed soon after? That was a different story when we planted twelve-inch-tall honey seedlings. It turns out that most folks would rather have a fairly low-maintenance system—where they have to wait a bit longer for shade but don't have to thin much—than plant heavily and have to thin heavily. It's also easier on the budget. The per-tree cost will go up, but the per-acre cost goes down.

One challenge when planting slow-growing trees like oaks or hickories is that they will need protection for much longer than faster species. You may want to remove protection from some trees, but be limited because you still need to protect their neighbors. That's where there can be a case for separating fast- and slow-growers—either in different rows or perhaps in different parts of the farm altogether. You could cluster all your oaks and hickories in their own groves, while the bulk of pastures feature faster species. That might also be useful for harvesting tree crops, because you can concentrate certain crops in their designated areas, keep livestock out, and harvest more efficiently. One layout we have done incorporates diversity rows every so often, which allows you to trial different tree species—including the slow-growers—into one management area that will need protecting longer.

There is a very common belief in the tree-planting world that a small seedling will eventually outgrow a much larger transplanted tree and so you should just go with the small seedling, which is going to be much cheaper, anyways. And yet, planting taller trees seems to make sense in this new context of silvopasture. It's only recently that I've come to formulate the reasons why it makes so much sense to go big when planting into pastures.

If I was planting into a context with no deer pressure, no cattle, no pigs, and no need for tree protection, I think I would indeed plant smaller trees, not much more than two feet tall. But when planting into a pasture, you need to protect that tree from damage, and that protection presents a unique environment. While tree tubes are super useful, trees don't grow naturally inside of them. They grow tall and lanky, unable to extend side branches until they reach the very top. The taller the tube, the stronger this effect, and since we use six-foot tubes for cattle silvopasture, the trees tend to emerge from the top as one long, gangly stick. Once the tree emerges from the top, it shoots out side branches, relieved to no longer be confined, which then makes the tree very top-heavy and certain to flop over if you were to remove the shelter. This means that the tube and stake are then required to keep the tree from falling over for another year or two once it has emerged from the tube, until the tree has the chance to develop the right caliper for its size.

Meanwhile, a tree grown out in the open setting of a nursery naturally develops a much healthier caliper to it. It is going to be much wider at the base, able to hold itself up naturally, and sway around the tube much less once planted. If I could plant only trees that were six to eight feet tall, I believe they would be much healthier and stronger, even if I still placed them within a tube for the first several years in the pasture. If the tree is tall enough to be above browse height, like seven or eight feet, you then only need to protect from rubbing, which could save considerably in tree shelter costs.

On another note, our survival rate is going to be higher when we plant tall trees. We are, quite simply, able to take much better care of a tree in the nursery than we can in a pasture. They grow taller, faster, with better caliper, and without the threat of browsing or rubbing. This also means that we can save considerable time and money on maintenance and tree replacement. Even if the survival rate is the same, skipping those first few years when the tree is small and subject to competition from weeds inside and around the tube and more prone to damage from rodent girdling will give you better results, with less time spent on post-planting care.

As we invest more in the genetics of our trees, we might as well invest in more care before we plant them. Grafting a honey locust costs the same whether you later grow it out to two feet or ten feet. A mulberry or persimmon grown from tissue culture costs way more than a seedling to start off, but the cost is the same to take that small tree and grow it out for another year or two. These trees represent real investments in the farm, and getting the same genetics in a taller package is just a better value.

In economic terms, having a quick turn-around from the time you invest in the tree to when it yields shade or feed for you makes a real difference and puts dollars in your pocket. And there's a real psychological benefit, which should not be underrated, in having trees that you can see from day one. If Uncle Bob or a farmer from down the road comes over to look at your new tree planting and sees only tubes, they'll have a hard time catching the vision for what you're creating. It'll even be hard for you to keep that vision alive, looking at white plastic without the sight of green. It makes a world of difference when you can see your trees standing proud and have to look up to admire them.

Most conservation plantings also don't receive nearly as much care as we give our silvopasture trees. If anything, they get an annual shot or two of herbicide, not the buckets of moisture-retaining, weed-suppressing, fungi-feeding mulch that we're so fond of. That goes a long way to ease the transplant shock of a new tree, even if it is large.

CHAPTER 14
Prioritization and Phases

It does not matter how slowly you go as long as you do not stop.
—Confucius

Now that you have a rough idea for how trees can benefit your farm, it's time to start thinking concretely about where you'll plant them. Since you'll want to avoid biting off more than you can chew, you'll want to break up the installation into multiple phases over the course of several years. The larger and more ambitious your silvopasture plans, the more time you'll want to give yourself to complete them.

What you'll need to do here is determine which areas should receive highest priority and which areas can wait a bit. A map of the farm is useful, and so is mulling this question over as you walk around or go about your chores. Here are some priorities past clients have set:

- Trees near the home that can be cared for and observed closely
- An orchard site for home use
- A pasture that is particularly hard to graze in the summer because of lack of shade
- A pasture you would like to subdivide into multiple paddocks using the tree shelters as fence posts
- Trees along cow lanes that will shade cows as they walk back and forth
- South-facing slopes that dry out early every year
- A really wet paddock you'd like to dry out
- Far paddocks where you don't want to travel every hour to move the shademobile
- An exposed site or heavy use area where you really want a windbreak
- Marginal ground that isn't likely to be anything other than pasture
- If a farm has pastures split into day and night pastures, the day pastures probably get priority

SAMPLE PLANTING PLAN FOR 100-ACRE GRAZING OPERATION:

Winter 1: Craft initial plan for farm. Each subsequent winter is for reflecting, adjusting the plan, and ordering materials for the next planting.

Spring 1: Plant ten to fifty trees. This is your training phase. The goal here is to get comfortable with the process and make any mistakes on a small scale so you can use those insights on a larger scale planting. Plant trees somewhere where you'll often see them so that they'll be on your mind.

Spring 2: Plant one hundred to five hundred trees. First sizable planting. Plant in the highest-priority areas, where trees are most needed. Focus on trees for livestock, which tend to be hardiest. If shade is your primary goal, consider going heavy on fast-growing species like willows, poplar, and black locusts. You can fill in with the slower-growing mast species in future plantings.

Spring 4: Plant the next five hundred to one thousand trees. Now that you have ample experience, you can tackle a pretty large planting with confidence that you know how to see it succeed. Work your way down the list of priority areas. By this point, you can likely start to re-use some materials from your first plantings.

Spring 6: Plant the next five hundred to one thousand trees. Now that you have several years of experience planting hardy trees for livestock, consider planting the best areas with marketable tree crops, if that fits your goals. Otherwise, keep adding trees for greater resilience and profit from your livestock operation.

This staggered approach to tree planting allows you to gain the insights, habits, and skills needed to get trees established well, to learn through experience which trees are right for your farm, and to give you time to adjust initial plans as needed.

The very first small planting is probably the most crucial. It gets some momentum established and will give you the confidence and know-how to get the rest done. A person can read everything that's been written about establishing trees, but the real learning comes through hands-on experience.

REASONS TO GO SLOW

- We're talking trees here. They grow slow. If you were growing salad greens, you could go through twelve generations of plants in a year and be an old hand in two years. In two years, some of your trees might barely be sticking out of their tubes and will be years away from bearing their first crop. So it takes time to learn. And there's a lot to learn about trees. This is a whole new element you're adding to the operation.

- The more complex the tree component you plant, the more you'll need to know about it. That learning doesn't happen overnight. You could try to learn everything you need to know about a system before ever putting a tree in, but you learn better through doing. And you'll never have it perfectly figured out anyways. So make plans, start small, plant some trees, learn, and reevaluate your original plans as needed.

- You will make mistakes. Expect that. The people who don't make mistakes are the ones who aren't trying anything interesting. What we want is to make mistakes on a small scale, rather than on a large planting.

- The trees you put in the ground will need to be cared for. If you put your whole farm into silvopasture at once, there's suddenly a lot more to manage. We can make it as low-maintenance as possible, but there will always be some maintenance. Going slow staggers the amount of time you'll need to devote to care. Please don't invest time and money into a tree planting that you don't have time to maintain. This is one major reason we prefer using tree tubes with electric fencing attached versus fencing out large areas, because the ongoing maintenance is much lower.

- Planning years in advance and building up towards that gives you much better funding options. Funding cycles for loans or NRCS grants are long, so the earlier you apply for funding, the better options you'll have to choose from.

- The selection of silvopasture-specific nursery stock is set to improve and increase significantly in the years to come. It will pay dividends to wait for better stock while working with trees where the genetics matter less or are already at a good place.

CALCULATING YOUR TREE NEEDS

Now that we've determined what trees you'll plant, how they will be laid out, and which areas you're first going to focus on, it's time to calculate how many trees you will need to start with.

You can calculate your tree needs based either on the acreage you'll be planting or on linear feet of the rows you'll plant. Using linear feet is the most accurate, but going by the acre allows you to get a good estimate.

When planting an entire pasture in trees at a regular spacing, you can calculate the number of trees you need based on the spacing between trees.

- Calculation is 43,560 square feet (one acre) divided by the row width, divided by the space between trees in the row.

- For example, say I want to plant my rows 60 feet apart and my trees will be planted 20 feet apart in the row. That means I'll need 36.3 trees per acre (43,560 ÷ 60 ÷ 20).

When calculating based on linear feet (say along a fence line or cattle path), simply divide the total length of the row by the distance between trees in the row.

- For example, a 1,200-foot windbreak with conifers planted every 10 feet will need 120 conifers (1,200 ÷ 10).

See the worksheet for more space to make calculations

CHAPTER 15: Aftercare

Of course, getting the trees in the ground is only the first step. The real goal is not to get them planted, but to have them mature to a stage where they can contribute to the farm. The first year is absolutely the most critical to tree survival—one more reason to not plant a ton of trees at once, since you don't want to be overwhelmed with aftercare. That said, good and thorough care in the planning and planting phases will make aftercare a breeze.

I can't stress enough how important it is to be checking in on your trees the first few years and fixing minor issues before they become major issues. Pulling weeds that come up inside the tube. Making small corrective pruning cuts if there are branches growing where they shouldn't be. It really doesn't take much time (averaging less than a minute per tree unless there's major issues), but it does need done. So get it on the schedule and on someone's to-do list.

A common question is how much water trees need. Generally, we do not water the trees we plant, yet I also live in a fairly forgiving climate that typically doesn't experience severe drought. The very best watering you can give is to apply a generous layer of mulch, which will conserve enough moisture in the soil for your young trees. Although mulch is bulky to apply, it is much lighter than water and more efficient than going out to water multiple times. The only time we water is when the soil is bone dry when we plant and there's no moisture to conserve with mulch. If time allows, it's great to mulch the places where trees will be planted several months in advance, which will trap moisture and start to build up the soil biology before the tree is even planted.

We do not use or recommend fertilizer at the time of planting, because we want the tree to focus on root development, not upward growth, the first year. Some fertilizer may be useful in later years, but not initially.

I am sure there is ample room to enhance the start of these trees through better agronomy, and I look forward to the day when we know as much about optimizing the growth of honey locust as we do about apples. We're just not there yet, though thankfully most of the trees we work with really are not fussy. Most of them are pioneer species, well-adapted to growing in conditions with strong competition and lots of herbivores. Just remember that in nature, you need thousands of seeds to get one mature tree, and we're trying to get all of our trees to survive. So we need to provide a little assistance.

REMOVING SHELTERS

The first thought many have after they've installed a tree shelter is, "When can I take this thing off?" The answer, of course, depends. The shelter provides several services:

- Protection from browsing
- Protection from rubbing
- Staked support while the tree is young and top-heavy

Once a tree has grown above the tree shelter, it no longer needs protection from browsing. However, these trees should continue being protected from rubbing until they are more mature—ideally until they start to grow too wide for the shelter. Even after they start to outgrow the tube, the bark can still be vulnerable to rubbing or stripping, especially because the tree was grown inside of a humid, protective tube that limited its opportunity to harden off. I learned this the hard way. I was trying to be proactive on one of our first sites and went to remove a bunch of tubes. All was fine during the winter, until we noticed that some of the trees, mostly willows, had been stripped of their bark in the spring. It seems that the spring sap flow makes the bark easier to peel and likely tastier. Spring is always the time to pay the closest attention to livestock wanting to strip tree bark. What we do now is cut the tubes open but leave them installed for a few more years so the bark can breathe and harden off while still being protected.

This is a black locust that was starting to bulge out of the tube so that fallen leaves were getting stuck in the bottom and rotting, which caused the bark to rot as well. The bark oozed water like a sponge when I pressed on it—not a good thing. Thankfully, the tree recovered nicely after we opened the tube up and let it breathe. This seems to be less of an issue with species other than black and honey locust.

Once the tree has the girth and strength to stand firmly on its own, you'll want to remove the stake. You can then reuse the stake—which is roughly half the value of the shelter system—in a future planting phase. If the stake is not removed, the tree and its roots will grow around it, making it much harder to remove. If the stake gets swallowed hard, you may need to cut the stake at ground level. The challenge with using regular T-post pullers or devices designed to pull small trees up by their roots is that they will snap the stakes in dramatic fashion, leaving long shards of fiberglass that you don't want to deal with. We've had some success using a vice grip and crowbar to loosen stuck stakes.

If there's still life in the tube when it's time to remove it (the manufacturer estimates they'll hold up for about six to ten years) and you have a second use in mind, you'll want to cut it in a way that makes it easy to reuse. We figure that tubes protected with an overpass configuration will probably just be used once, since the tube needs to stay in place longer to provide the height for the top wire and insulate the tree from the spiral wire. Single strand configurations can potentially see the shelter removed in just a few years, especially if the tree is larger to begin with.

Once a tree starts to grow too big for its tube or doesn't need the full protection of the tube anymore, the tube can be removed and replaced with this if protection from rubbing is still required. Note that the tube has been cut but is held together by a wire, which will give it space to expand for several more years.

Without a stake to hold it up, the wind pushed this tree over once it leafed out. Since it had grown up inside of the tube, it had hardly any girth and was extremely top-heavy—which is the rule, not the exception, if trees start off small inside the tubes.

Be careful not to remove stakes too early. One winter, I decided I would pull fiberglass stakes from an older project so we could reuse them elsewhere. I pulled several hundred, and it didn't seem to affect the trees at all—except that many of those trees started to lean, some all the way to the ground, and all in the direction of the prevailing wind. Once the leaves came out, the trees became top-heavy masts with hundreds of tiny sails catching the wind. Since the trees had grown up constrained inside of the tree tubes, they didn't have the appropriate girth to support the tall, lanky trees or the tree tubes around them. So I had to go back and re-stake hundreds of trees again. I've been much more patient in removing tree tubes ever since.

One issue we've seen with tree shelters is that there's a tendency for the tree to be damaged by rubbing up against the hard shelter surface day after day. This is worse with some species than others. Black locust seem to get it the worst, and my hypothesis is that it's caused by their vigorous outward growth once they escape the confines of their tube. This makes for a very top-heavy tree that sways heavily in the wind. A simple fix is to make small cuts in the top of the shelter so there's more give and less of a sharp edge to rub against (one-half to one-inch cuts all around and then folded down to make a flower shape). We have also used rubber ties—commonly used in the vineyard industry—to tie the tree to the top of the stake, which limits the back-and-forth motion of the tree, although it also seems to limit the ability of the trunk to develop caliper as quickly. Plantra is also actively working on a solution to this problem by developing a gentler top.

Cutting away at the top of the tube can help you avoid abrasion like this.

Minimizing aftercare needs is a major reason that we're moving towards planting larger and larger trees. I would love it if we could get to the point where all of our trees are sticking out from the tube the moment we plant them. There's not much instant gratification in the world of tree planting, but that would be one great example. We can just take so much better care of trees in a nursery than we can in a pasture. Those grown to be six feet tall or larger in a nursery have a better caliper, because they haven't been confined by a tube. They won't get whipped around by the wind as much when they emerge from the tube. They won't get overshadowed by weeds, either inside or out of the tube. They should mature more uniformly and with higher success rates than little seedlings. We pay more for the trees on the front end, but the consistent quality and maintenance savings more than make up for it.

Securing the tree to the stake like this will reduce rubbing significantly. It does also seem to reduce the strength of the trunk, because it isn't getting "exercised" by moving back and forth as much. These are some of the challenges when working with tree tubes and reasons to transplant taller trees when possible.

Lastly, some trees will die, no matter how well you care for them, so you'll most likely have to go back and replant some portion of them. We'll often wait two years to replace the trees, which gives us two growing seasons to see what makes it and what kicks the bucket. Also, some trees that look like toast in the first year will actually come back. This is especially common for persimmons. I once planted a row of hundreds of tiny persimmon seedlings to later graft onto, and the mortality was so high I thought I lost 80 percent or more of them. But by June of the next year, they came back from their roots with a vengeance—so much so that I probably ended up with a 95 percent success rate. Had I replanted after the first year, I would have missed all those persimmons coming back.

PRUNING, POLLARDING, THINNING

Because we're aiming to create a system with the ideal balance of shade and sunlight and that balance will constantly shift as the trees age, you will need to manage the trees over time through pruning, pollarding, and thinning.

Pruning. One of the attributes of a well-managed ruminant silvopasture is that the tree canopies start high up on the tree, so the shade moves significantly throughout the course of the day. Pruning lower branches in the early years, while these branches are still small, is a quick and easy task. In later years, pole pruners can cut small-diameter branches from a height, or a pole saw can remove larger branches as needed. Proper and timely pruning of trees for the purpose of letting in more light will also create a higher-quality saw log for future timber use. Trees prefer it when you prune them in the winter, when their energy stores are in their roots, but if you want to use prunings as leaf fodder, that needs to happen during the growing season, which is also fine for the trees. Pruning in late summer, when the tree is not actively growing, minimizes damage and tends to coincide with when forages are at their lowest point.

There is a balance to be struck between keeping all the limbs in the first years to maximize precious shade and moving the canopy up. We have gotten a bit less aggressive in pruning really early so that the trees can provide more shade when they are young. Also, if you prune too high up on young trees, they tend to flop from being top-heavy with not enough caliper to support them. So start pruning slowly, figure out what works, and keep at it year after year.

Pollarding. For willows, poplars, and mulberries in particular, pollarding for leaf fodder will provide valuable summer feed while keeping the tree small and manageable. Once the slower-growing trees are casting enough shade for your livestock, the faster-growers can be pollarded more regularly for use as leaf fodder, which also reduces the amount of shading on the pasture.

Thinning. As the tree system matures, some trees may eventually need thinned out—to create more space for nearby trees, allow more sunlight in, harvest for timber, or a combination of the above. This will eventually create an abundance of wood products for farm use in the form of building materials, fence posts, firewood, mushroom logs, etc. You will be setting the farm up to supply many more of your own needs than you currently can. Many, though, will opt to plant at lower densities and skip thinning altogether, accepting a slower rate of establishment in favor of lower costs and less labor per acre.

High rates of failure may have been the rule in pasture tree planting throughout most of history, but it doesn't need to be going forward, especially since you now have a cheat sheet. We at Trees For Graziers now have so much confidence in our system that if we're involved in the planning, planting, and maintenance of trees, we guarantee our clients a 90 percent success rate, and we'll replace anything that falls beneath that threshold in the first year. That confidence comes from having established simple, replicable protocols that work time and time again. We recognize that the main killers of otherwise healthy trees are livestock, deer, rodents, and competing vegetation and have a plan that addresses each one of those factors. We take simple steps to make sure the trees get the best care, like keeping the roots cool and moist and giving them a generous helping of mulch. All of this adds up to a high rate of success, which you can use on your farm as well.

IF YOU ALREADY HAVE TREES

So far, we have focused only on the addition of trees to open pastures. That is what I have the most experience with in southeastern Pennsylvania, which has been farmed and grazed heavily for hundreds of years and where land typically sells for well over $30,000 an acre. Not many people let that prime farmland grow up into woods. Also, by adding trees to open pastures, we have exact control over what species we'll use, which genetics we select, and where the trees are placed. There's no dodging stumps or logs, and you can easily access pastures to clip them if you need to get some weeds under control.

That said, many regions where grazing thrives are only marginally agricultural, and woodlands dominate or encroach on the edges of pasture. Thinning the woods to create pasture gives a much quicker return on investment than waiting for planted trees to give you the shade and feed you want. So I wanted to include at least a bit of information here about the upsides of thinned silvopasture.

The most important step here is to learn to recognize what you already have on your farm. It's amazing how much value we can see around us when we look through new eyes. For instance, one homestead I visited ran a small flock of sheep and wanted shade as well as shelter for the cold months. The plan was to plant white pines as a windbreak, as well as build a barn for shelter, while also thinning trees out of a low-lying section of the farm. Yet as we walked around, it became clear that a lot could be done by simply working with the existing vegetation. The red cedars that come up in so many pastures had formed a dense and well-protected area that could shelter from any storm. Better yet, there were honey locust trees growing above the cedar grove, dropping pods down for the sheep to browse during the winter. They weren't thornless or great pod producers, but they were out of the way and already contributing to the farm. The autumn olives could be cut as fodder and the densest tree areas thinned over time to allow more light to reach the ground, without sacrificing the windbreak effect of the cedars.

Each area seems to have a handful of species that readily emerge in pastures. Black walnuts and mulberries are prolific in southeastern Pennsylvania, while pasture walks in Virginia almost always show persimmons. All of these are very useful trees in the right context, and it's often easier to protect them than try to plant a new tree. If you can graft an improved cultivar to a volunteer seedling, even better.

Educating yourself is the best tool here. Learn to ID the plants coming up already, and ask yourself, "How could I make use of this?" Learn how to prune and shape timber species so they produce a nice saw log, see what invasive shrubs your livestock will gladly eat if you cut branches for them, and try protecting trees that are already coming up in your pasture. See if you can pile the branches you just cut from the thorny, invasive shrubs into a circle around that little persimmon sapling coming up, and maybe even graft it a few years down the road. This might not get trees everywhere on your farm or in the most systematic way, but it's a great, free way to start gaining new skills and insights while adding to the usefulness of your farm.

CASE STUDY
Wild Harmony Farm

Ben Coerper of Wild Harmony Farm in Rhode Island farms a tract that, though large by Rhode Island standards, is almost entirely wooded. As such, he was extremely limited with the space that could actually be grazed. Access to shade wasn't the issue. Having enough sunlight to grow grass was, so it was time to enlist the chainsaw.

Ben was able to partner with a local logging company to thin twenty-five acres over the course of six months, with no money exchanged in the process. The company also agreed to pile the tops of the trees. The goal was to end up with a 50 percent canopy that favored the mast-producing species like oaks and hickories while keeping a smattering of black gum, tulip polar, white birch, and black locust for diversity.

This simple technique can save thousands of dollars on even a small fencing project. Photo courtesy of Ben Coerper.

The next step was to fence the new silvopasture in. Anyone knows that if you mount a fence right to a tree, that tree will eventually swallow the hardware whole. But if you nail a board to a tree with a long nail and a wide washer, you can then attach your fencing hardware directly to the board, saving many thousands of dollars compared to establishing traditional fencing.

Note the wide washers. Those are essential! If you don't include those, the nail will pull through the board, undoing your work. Also, consider using long-lasting boards while you're at it.

Forage establishment was done through a combination of broadcast seeding (good luck getting a drill in there) ten pounds per acre of orchardgrass, annual ryegrass, perennial ryegrass, and tall fescue in various experimental plots. If he were to go back, he would have used twenty to twenty-five pounds per acre of a mix of orchardgrass and annual ryegrass. Winter bale grazing also added a lot of seed, and many graziers note that it can be the best way to seed and kickstart the biology in a new section.

What gets especially interesting is the economics of this conversion. In a region where open ground goes for $10,000 an acre and wooded land for more like $3,000 an acre, starting with woods and thinning them gives you a wide margin for error. In the case of Wild Harmony, the logging was free, seeding was $80 per acre, and installing a two-strand high-tensile wire cost only $60 per acre because of savings on fence post installation. $140 per acre is a bargain!

Given the positive experience, Ben is bullish on the future of thinned silvopasture in the region, since New England has only 9 percent of its land in ag, versus 67 percent in forest. As he says, the region grows three things well: trees, rocks, and grass!

Thinned silvopasture at Wild Harmony Farm. Photo courtesy of Ben Coerper.

We at Trees For Graziers see a real need for foresters who can consult on thinned silvopasture in the coming years. Thinning the woods is something most farmers have little to no experience with, and once you've cut a tree, you can't just glue it back in place. Meanwhile, few foresters have experience with livestock and forage establishment, and many are hostile towards the idea of thinning woods for cattle. Yet as forester and grazier Brett Chedzoy of Cornell Extension says, once you've kept the high-quality timber trees, the next tier will either grow you firewood or grass, and grass is a lot more profitable than firewood. We need more foresters like Brett who can help landowners thoughtfully decide on what stays, what goes, how to make use of the tops, how to make money (or pay as little as possible) on a thinning, how to negotiate bids and terms with logging companies, etc. It's a high-stakes game, which is tough to do when you're a farmer playing this for the first time. That's why an experienced forester is so valuable to the process. This is a real untapped business opportunity, helping farmers throughout your region expand their grazing and increase their resiliency.

SECTION 4
Funding It

I believe that if we could plant a tree and have it mature and yield a crop in a year, we would already have silvopasture everywhere. But the fundamental challenge of establishing trees on farms is that it takes years and sometimes decades for your investment to pay off, whether that investment comes in the form of sweat equity, money, or a combination of the two. While farms are businesses that are used to paying large sums of money—for land, livestock, barns, machinery, and more—their solvency tends to require that they get cash flow from those investments pretty quickly, or at least make enough for bank payments. While a silvopasture system for a hundred-acre farm might not cost more than a new tractor, you get to use that tractor right away, not five years after you buy it. This dilemma requires a bit of creativity.

The primary way we have gotten silvopasture systems going in our region of Pennsylvania to date is by finding grants to make them free, or close to free. Without that, we would only have able to do a small fraction of the projects we've completed, especially since our earliest projects were full of trial and error. While I don't want to rely solely on grants in the long term, they were crucial in allowing us and our clients to take a risk with an unproven concept for which there were very few guides. Now that we have, as of 2025, about six years' worth of experience planting trees in pastures and protocols that consistently yield good results, we are seeing more folks willing to put their own money on the table. Yet even if you're confident silvopasture will work and end up being profitable in the future, you still must cashflow it.

Up to this point, I have been highlighting the benefits of silvopasture to the farm operation itself. But as you can imagine, there are a whole host of ecological benefits that come from planting trees into a pasture to create a more diverse, savanna-type ecosystem. Trees will create more habitat for birds, mammals, and a whole array of insects and pollinators. They will take carbon dioxide from the air and lock it up, both in living tissue and soil carbon. That soil carbon will allow the soil to capture more rain, which will reduce flooding in storm events. All of these ecosystem services are beneficial not just to the farm itself, but also to those outside the farm—which is why there are a whole host of groups who financially support tree planting. Since you will be providing tremendous ecosystem services to your community, using outside funding is a great way for the community to acknowledge the value of your farming practices and support your work. I can think of few better places to spend taxpayer dollars than planting trees in pastures in a way that makes regenerative farms more viable and makes for a healthier landscape for the long term.

Robins in a pollarded hybrid poplar.

CHAPTER 16
How to Find Funding

NRCS FUNDING

When it comes to getting outside funding support for silvopasture, the first place to look is your local Natural Resources Conservation Service (NRCS) office. NRCS is specifically built to help farmers throughout the country with conservation practices. The problem is, planting trees, especially hardwood trees, into open pastures has hardly been done in the US. Because there's been little demand for planted silvopasture, many states do not fund it. On top of that, funding rates vary wildly from state to state. In 2021 in Virginia, the subcategory described as "Establish hardwood trees in an existing pasture with adequate forage" got cost-shared at $369.87 per acre. Georgia had less precise subcategories and offered $139.71 per acre for "Tree Establishment." Meanwhile, Missouri offered $25.44 per tree for "Bareroot Trees and Shrubs with Tree Protection" and confusingly offers a mere $5.80 per tree for "Bareroot Trees and Shrubs, with Tree Shelters." While more states have started to offer funding for silvopasture since the first edition of this book came out and others have increased their rates, it's still a patchwork quilt of rules and rates, and few NRCS staff have any experience with the practice.

If your state recognizes silvopasture as a fundable practice, it can fund adoption through either EQIP (Environmental Quality Incentives Program) or CSP (Conservation Stewardship Program). The major difference between the two programs is the way they pay out their cost-share. EQIP pays a larger amount soon after the project is completed, while CSP pays a tiny amount initially and the bulk of the money is paid out in the form of annual payments, of which the minimum was increased in 2024 from $1,500 to $4,000 per year over the course of five years.

When it's time to meet with your local NRCS agent to apply for silvopasture funding, you'll want to bring along the plan you have developed, as well as the NRCS silvopasture sheet included in the appendix. Most NRCS staff will be unfamiliar with silvopasture, so you may need to educate them, and having a document developed by NRCS itself might reassure them this isn't some harebrained thing you're doing.

Tip: CSP can be used for silvopasture, regardless of whether your state pays for silvopasture. Here's how it goes:

- Apply for CSP for help with a conservation practice. That could be silvopasture or any number of other practices, some of which are very inexpensive to implement.

- Sign the contract, establish the conservation practice, and receive the initial small cost-share, plus annual payment of $4,000 or more.

- Use that payment to purchase silvopasture materials. By providing your own labor, you can keep costs low and fund over a hundred trees per year at that rate.

- This "CSP Strategy" can work well for small farms wanting to slowly and steadily implement silvopasture on a tight budget, regardless of state.

For those looking to convert cropped or hay land to pasture, or who want to integrate trees on the borders of their fields, alley cropping, windbreaks, and hedgerows are all NRCS practices to consider. They can, in certain instances, give significantly better cost-share than the practice of silvopasture. One will hardly be worth the application; the other might cover all your costs. It's worth looking into. Or, if silvopasture is not offered but something like windbreaks are, maybe you start your tree-planting endeavors with windbreaks instead.

Do keep in mind that sometimes it's just not worth jumping through whatever hoops NRCS puts up in order to qualify for the money. It's even worse when you'd need to drastically change your plan to fit within the box that your local office might enforce. I have heard of several folks being turned down from silvopasture funding because their plans included mulberry trees, which are deemed invasive. This, despite the fact that both farmers were in Iowa, surrounded by monocultural ecological deserts of corn and soybeans. If they can take NRCS money for the other tree species and add mulberries on their own, that might be a good compromise. But don't give up on a project or leave out the aspects that will do the most good for your farm if a particular funder won't approve it at a particular time. Go find another way to make it happen.

CONSERVATION AND FARM GROUPS

Every region has local conservation-minded groups who are more than happy to support tree plantings. These groups will differ by region, and you'll need to reach out to find them. In our region of southeastern Pennsylvania, conservation groups that have supported tree planting include the Chesapeake Bay Foundation, the Alliance for the Chesapeake Bay, Trout Unlimited, Stroud Water Research, and many more. If you're connected to a farm group like Practical Farmers of Iowa or a large brand like Organic Valley, reach out to them to see if they're aware of local resources that can help you. Some states also have significant state-level ag conservation programs to look into. Here again, taking a patient approach will pay dividends. Many groups want to support tree planting and will readily get behind it in support, but funding cycles take several years, and they may not have any funding in hand that can be used for silvopasture. Given the time, many groups would willingly go after funding to help you out, if you take the time to work with them.

Because NRCS has, to date, been unwilling to provide financial support for silvopasture in Pennsylvania, we've mostly relied on conservation nonprofits to fund our projects. In general, they are going to be significantly more flexible than the NRCS, though funding might be less consistent. We have had to work within the constraints of whatever funding a nonprofit partner could acquire or reconfigure for a silvopasture project, and we've had to work with a different group just about every year because we were mostly using one-and-done grants—or putting expiring funds they needed used toward one of our plantings. That makes long-range planning challenging, but it's helpful to put yourself on their radar in case they need to use an expiring grant.

LOANS

In a context where outside grant funding is too small, too restrictive, or both, loans make a lot of sense. They are specifically built for purchases where there's a large up-front cost and a much longer payback horizon—whether that's a farm, barn, or trees. Using a loan has the great upside of reducing up-front costs and cash flow pinches and allows you to pay off the trees over time as you're earning more profit from the trees.

The challenges with pursuing standalone loans for tree planting projects include scale and the inability for a bank to take a lien on a tree. Unless your planting is of significant size (say over $20,000), it would generally take a high interest rate and up-front fees for a bank to make it worth their time. The beauty of tree planting is that it's an investment in your farm that is much more affordable than buying a barn or tractor or land. Yet a bank prefers big, expensive purchases that they can take a lien against and resell if you default. If you don't repay your tree planting after five years, nobody is going to dig up and repossess the trees. That said, we have talked with several rural banks who are used to working with small farms, and all seemed quite willing to fund silvopasture if there was equity on the farm or even a herd of cattle to lend against.

Silvopasture emerges as a very worthwhile investment when compared to increasing production through buying land. With the price of land increasing faster than the price of what you can produce from that land, it will be more and more difficult to profitably increase production by expanding your land base. Increasing production by adding trees should be hands-down the more profitable option in most areas, even without factoring in cost share or the increased production efficiency from shade and lowered heat stress. Just looking at the additional feed coming from a well-managed tree system (feed that's there to replace expensive cow-days of July or January, not the cheap cow-days of May or June), you'd be hard-pressed to match that kind of additional feed by buying new land—assuming you even could find land close enough to make it worthwhile. If land isn't adjacent, you start to really ramp up the cost of production due to trailering, more work to monitor animals further away, and problems that get missed when you can't keep a close eye on stock.

This seedling honey locust started yielding in its fifth season after planting. I believe we can speed that up considerably with taller stock and better care than this tree was initially given, but it gives an idea of the wait time for a return on investment.

Let's go back to a previous example where we used a conservative yield estimate of 100 pounds of pods from a mature honey locust. If we assume it costs $50 to establish that tree, then our annual loan payment is $4.30 over twenty years at 6 percent interest. If we're again conservative and assume those 100 pounds of pods will give us three cow-days, then we have a cost of $1.43 per cow per day. In reality, you would almost certainly get more feed value from the pods, because they contain roughly twice as many calories per pound as hay. Let's compare these calculations to the cost of gaining production by acquiring new land. We previously used the example of Virginia hay ground producing 2.5 tons of forages annually, which at 33 pounds per head gives you 152 cow-days. In this case, that's an optimistic number, given that we should be accounting for wastage and trample. Even still, to match the $1.43 per cow-day price with the same loan rates, you'd have to find land costing less than $2,527 per acre. I don't know about you, but there's no land at that price in my area.

Unlike land, you won't have to pay annual taxes on the trees. And remember, the feed you're getting from pasture forages tends to come when you already have plenty of forages. The main point of adding trees for feed is so that you can have feed when forages aren't growing. Since land is seen as a place to invest and store wealth, the cost of land tends to be completely disconnected from the profit potential of that land. The competition for land from folks who have off-farm jobs and just want a nice place in the country to have a couple cows and hunt is a huge driver of the cost of land, and those buyers care very little about the productivity of that land. So if you want to store wealth or have a nice place to call your own, buy land. But if you're serious about making a profit from grazing, investing in the production potential of your current land through silvopasture is likely a better way to go. As Allan Nation says:

> For the first time in North American history, the average buyer of farm and ranch land is not expecting any production income from it. Its whole value is tied up in aesthetics and lifestyle. In this new economy, an occasional glimpse of wildlife is more valuable than any amount of livestock carrying capacity and a pretty pasture is worth double that of cropped land. The productivity and net return of land is no longer a major consideration in what today's land buyers are willing to pay for it. Unfortunately, most agriculturists are still carrying in their heads the old paradigm of paying for land from its production. This has led to huge financial stress.[1]

PAYMENTS FOR ECOSYSTEM SERVICES

The recent buzz is for farmers to get paid for ecosystem services like carbon sequestration. Unfortunately, I don't see that being very profitable in the next few years for all but the largest farms. There's a certain overhead required to onboard, monitor, report, and transact these ecosystem services, and it takes some large economies of scale to make it worth the while. This is the type of thing that makes much more sense in Brazil than in the US, since Brazil can grow trees much faster, cheaper, and at a much larger scale. We may have 10,000-acre ranches in the US, but they are in the West where trees hardly grow. Brazil has huge ranches in the tropics, which can grow trees faster than the lushest parts of the States. And scale matters in the carbon business, since folks paying for sequestration have huge targets to hit. It's not a game that small farms have any advantage in.

It's essential that these silvopasture systems be profitable without needing payment for ecosystem services. Payments for carbon sequestration, flood mitigation, wildlife and pollinator habitat, etc. are only the icing on the cake. The payment numbers for ecosystem services are also not all that impressive when compared to the increased production from a good silvopasture system. Just look at the value of honey locust pods alone, which can yield at fifty to sixty bushels of corn equivalent per acre per year. If we assume even half of that yield, thirty bushels per acre, and assume organic corn is valued at $10 per bushel, we're talking about $300 per acre per year. Carbon payments, on the other hand, if paid

1 Allan Nation, *Land, Livestock and Life* (Green Park, 2007).

out at the current standard of $15 per ton, might yield $15–75 per acre per year, depending on your rate of sequestration and who is making the calculations. And that doesn't even take into account the administrative fees that would be taken out so you can sell your carbon offsets to Microsoft or Amazon.

Carbon offsets and other ecosystem services may be the hot new thing, and they can be quite useful if they can cover a significant part of the up-front costs for tree planting without being a significant monitoring and verification burden on the back end, but they should not be regarded as a magic bullet.

SCALE

When it comes to seeking funding support for a silvopasture planting, scale matters. If you're just starting out and want to try planting ten trees, there's no way it's worth your time and energy to knock on doors for funding support. If, on the other hand, you'd like to plant five thousand trees over the course of several years, it should be well worth your time to make some phone calls and see if you can cut the cost down a bit.

Between those two scenarios there can be an in-between where it takes considerable time and hurdles to pursue outside funding, but the payback for your time isn't great. Where that gray zone is depends on how easy it is to access funding, how much you have to spend, etc. Say you'd like to plant two hundred trees and you can get a local group to provide you with some free trees and shelters, but they're not the right trees (a bunch of maples for instance) and not the right shelters. You're much better off in that scenario paying for everything out of pocket than taking the free stuff that isn't actually what you want. You can, of course, also approach that group and discuss whether they can supply what you need in the future.

REDUCING COSTS AND SPEEDING UP RETURNS

It's worth repeating that the more you're willing to do your silvopasture planting over a longer period of time, the more you can decrease the cost of your planting. For instance, stakes and tubes can be bought the first time and reused several years afterwards, cutting the cost of shelters in half. You can buy willows and poplars initially, then use them to propagate future generations for free. And you can protect the trees already coming up in your pastures. Better yet, start your own consulting and tree planting company and service your area during your slow season. Then you get to buy materials in bulk, save money, and help others in the process. Anyone willing to get creative and roll up their sleeves can make silvopasture happen on their farm, regardless of available funding.

In the end, planting trees is an investment in your farm that will repay itself many times over. However you do it, however you get it funded, the important thing is this: Just make it happen. Every year you don't, you're leaving money on the table.

The other factor in cashflowing an investment is how fast it gives you a return. A $50 investment in a tree is a lot more expensive if it dies first, needs replanted, gets neglected, and only starts giving you shade in ten years, versus three if it is well cared for. Even if an oak tree were to have the same yield as a honey locust, if it takes twice as long to yield, it's harder to cashflow. That's also why it's so important to have the right genetics for certain trees. If each tree costs $50 to establish but only half of the trees yield fruits, then you end up paying $100 per productive tree. If a tree is capable of giving you $20 per year of return, then every year you can save between now and that yield will make you an additional $20. In that way, paying more for a tree that is one to two years ahead of a cheaper, smaller tree will quickly repay itself.

This hybrid poplar is in its second growing season and is already providing some shade—and it was started as a smaller pole than what we usually use now. Note that the mesh tubes were experimental, and we don't recommend that folks use them because cows are likely to get a taste for the tree when small branches grow out.

I'll lay out an overly simplistic example to drive home the importance of considering the time value of money. Imagine you could establish a mulberry tree that would yield you $30 annually. It would cost $50 if you planted a small tree or $60 if you planted a large tree. The small tree would hit peak production at ten years, and the large tree would hit peak production earlier, at only eight years. Using a conservative discount rate of 5 percent, the larger tree will net you an additional $18.33 compared to the smaller tree because it will begin producing earlier.

However, if you run the same numbers for a small and a large oak tree and assume they will reach peak yields at ages twenty-five and twenty-three, you get a very different story. Because of the much longer wait and the opportunity cost of what that money could have yielded you elsewhere, using the larger tree in this case loses you $1.76. Yielding quickly offers major advantages.

This is a reason we're very interested in dwarf mulberries, for example. Within a year or two of planting, they hit their max height of about twelve feet and yield just as quickly. They should make a really good fit for poultry systems, where they are quick to provide a thick, low cover for birds and then double as browse for ruminants. They might not ever produce as much per tree as a larger mulberry, but they are multifunctional and fast. Using big willow and poplar poles is similar. They may never give the range of benefits a honey locust gives, but if they can offer quick, cheap shade, that's a great option.

It's not much, but this dwarf everbearing mulberry started yielding about fourteen months after it was planted as a twelve-inch-tall leftover tree. Had it been planted as a five-foot tree, it likely would have yielded the same year.

It's tempting to look at a seedling tree that costs $3 and a grafted cultivar that's $30, think, "I could plant ten seedlings for the price of one grafted tree," and forget about all the other costs of planting a tree, including the opportunity costs.

Both a seedling or grafted tree will cost the same to protect, lay out, plant, mulch, and maintain, though a larger tree will likely need less maintenance. So the cost of the total package might be $43 in one case and $73 in the other. The difference is still considerable, but not ten times. And if the cultivar is a female tree that will guarantee large quantities of quality fruit at a dependable time of year, you're probably at least doubling your value compared to a seedling. In essence, we should be less concerned about the price per tree than about how much it takes to get a certain outcome. I don't want a pine plantation that costs me only one dollar per tree if it leaves me with a timber stand that I might someday have to pay someone to remove! If you stand to reduce your feed costs by $2,000 per acre by planting mulberry cultivars but go with seedlings instead and max out at a savings of $1,000 per acre, you've left the $1,000 annual savings on the table. It may have saved money on the front end but capped your upside significantly; spending a bit more to get the higher return would have been well worth the investment.

CASE STUDY
Rising Locust Farm

The folks at Rising Locust Farm have been bitten hard by the tree-planting bug. They are several years into the process of establishing a wide variety of trees throughout their diverse, multi-species grazing systems. Between an orchard for plums, pawpaws, hazelnuts, and a whole host of other fruits and nuts, plus trees for shade, nitrogen fixation, leaf fodder, and dropped feed, they have more trees than you can shake a stick at. They've taken an incremental approach, adding more trees bit by bit every year instead of tackling everything they might want to do at once. To do so, they've made use of a lot of different funding streams throughout the years. Some of that has been federal money, some has come through the state of Pennsylvania, and some through regional groups like the Chesapeake Bay Foundation. Like anything, they've found increased success with both tree survival and getting access to funding as they've gained experience. Harrison Rhodes of Rising Locust Farm says:

As far as the funding goes, we have used a single five-year Conservation Stewardship Program contract ($1,500 minimum annual payment for Haney Soil Test and timing our calving with the natural flow of the seasons), hundreds of free Chesapeake Bay Foundation trees and shelters, and the PA DCNR grant you know about, $55,000 for eleven acres [Trees For Graziers helped them acquire the grant, which was connected to stream conservation]. We have grown probably around 1,000 trees (mostly willow and poplar cuttings but also black locust and honey locust from seed) and had mostly volunteers plant the trees on our farm via tree planting parties (work them for 3–4 hours after a demo and then provide lunch, preferably pulled pork).

Closing

Here is something that I never get sick of. I have lived on this farm for 22 years and it is just getting better and better. It gives us a great feeling every time we come up here. It's a beautiful landscape, and it's improving every year. How do you value that? How do you quantify that? People try and design a good life in many different ways, and often it revolves around income. But you know, improving the landscape value gives a better feeling about where you live. It's not something that you can value with money, it's just a feeling. It's where you live, and if it improves your quality of life, it's a great thing.

—Noel Passalaqua, quoted by Rowan Reid in *Heartwood*

Silvopasture will make grazing into a force to be reckoned with. A quiet force, and a patient force, but one able and equipped to reshape the American farm landscape.

Hundreds of millions of acres of land are currently ecological deserts, their soil being monocropped to death, mostly to feed livestock that are crowded into inhumane barns and feedlots judiciously hidden away from the public. Grazing has been an answer to that, but only a partial answer. Grasses do not provide enough easily digestible energy for pigs and poultry, so they have been dependent on annual grains, even when supplemented with pasture. Cattle and sheep thrive on lush spring grasses, but struggle during the summer when temperatures rise and forages wither. Winter may bring relief from the heat, but it is also the most expensive time of year for graziers because they need to provide supplemental feed. And in all of this, grass farmers have had precious little protection from the vagaries of weather—too hot or too cold, too dry or too wet.

Trees are a game-changer—but not just any trees planted anywhere. Strategic, thoughtfully used trees, of the right species and genetics, will change agriculture, making it healthier, more resilient, more profitable, and so much nicer to work with.

This will not happen overnight. It will take patience and steadfastness from everyone involved. This is fundamentally a slow game, played out over the course of decades. But those with the knowhow and patience to use trees can rewrite the rules to the game and choose to play a game they can win. There will be tens of millions of acres of marginal crop ground that will now be more profitable to graze because of the addition of trees. Folks raising pigs and poultry on pasture will still need grain, but less of it, reducing the need for those crops even more. This is a symbiosis that has taken place at scale in some places and at some times in the past, but not in our current agricultural system. And nothing excites me more than seeing it play out on a larger and larger scale right now.

Lastly, what happens to farm families when kids get to participate in planting trees, then watch as what they planted grows up to tower over them, cooling them when they move cows or feed the chickens? What happens when there's hundreds or thousands of symbols of hope growing all across the farm, assuring you that next year, this farm will be better than it is now? That is a farm that will continue, because each successive generation has room to find joy and purpose on the farm. There are multiple incomes where previously there was none. There is security and resilience, when previously it seemed each year was one missed rain away from disaster.

This apple was the literal first fruits of one of my first ever silvo-plantings, and I couldn't help but tear up when they gave it to me.

I am excited and humbled to play a small part in this next phase of agriculture. But you are the ones to really make it happen. Thank you for your work in caring for the land, restoring your piece of this world, and showing that a better way is possible. Keep it going, and let's see just how regenerative grazing can be.

APPENDIX A

Silvopasture Article Reprints

1. A. A. Wilson, "Browse Agroforestry Using Honeylocust," *Forestry Chronicle* 67, no. 3 (June 1991): 232–35. https://doi.org/10.5558/tfc67232-3
2. O. A. Atkins, "Yield and Sugar Content of Selected Thornless Honey Locusts," Alabama Agricultural Experiment Station.
3. J. C. Grimes, "Value of Shelter for Wintering Beef Breeding Cows," Alabama Agricultural Experiment Station.
4. J. Russell Smith, "Chapter X: A Summer Pasture Tree for Swine and Poultry—The Mulberry," from *Tree Crops: A Permanent Agriculture* (Harcourt, Brace, 1929), 83–93.
5. George Washington Carver, "Feeding Acorn" (Tuskegee Normal and Industrial Institute, 1898).
6. USDA NRCS, "Conservation Practice Fact Sheet: Silvopasture — 381: Establishing a Silvopasture System," December 2019. https://efotg.sc.egov.usda.gov/api/CPSFile/15213/381_MD_GD_Silvopasture_Establishment_Fact_Sheet_2019_pdf
7. Leaf Nutrient Analysis
8. Honey Locust Analysis

Photo courtesy of Costa Boutsikaris.

1. Browse Agroforestry Using Honeylocust

Browse Agroforestry Using Honeylocust

by A.A. Wilson[1]

Abstract

The efficiency of establishing honeylocust (*Gleditsia triacanthos L.*) trees in operating pastures is being tested at the Springtree Agroforestry Project. Existing electric fences are used as fence rows and to provide protection for honeylocust which produce high nutrient pods for animal consumption. High production cultivars have been grafted to seedling nursery stock and planted out. Annual pod production can be self-harvested by sheep and cattle as a supplementary feed source. When properly spaced, honeylocust do not significantly reduce understory grass production. The literature on honeylocust as an agroforestry species is reviewed, and is used to develop financial rates of return for the program. Potential internal rates of return for pasture honeylocust plantings, calculated using a variety of production and cost assumptions, show net gains ranging from 9% to 24%. The formation of the Honeylocust Research Group is described and future research needs discussed.

Résumé

Au "Springtree Agroforestry Project" on est en train de mettre à l'essai l'efficacité de la plantation de gleditschias (*Gleditsia triacanthos L.*) dans des pâturages exploités. Des clôtures électriques déjà en place établissent les périmètres du terrain et protègent les arbres, qui produisent des cosses de haute valeur nutritive mangées par le bétail. Des variétés cultivées de haut rendement ont été greffées sur des souches de semis cultivés en pépinière, qu'on a plantées par la suite en plein champ. Les moutons et les bovins récoltent les cosses automatiquement et ont accès ainsi à une source supplémentaire d'alimentation. Espacés convenablement, les gleditschias ne réduisent pas de façon significative la croissance de l'herbe sous-jacente. Dans le cadre de l'article, on examine aussi la littérature sur le gleditschia comme espèce agroforestière, et on utilise cette littérature pour développer des taux de rapport financier pour l'opération. Il s'avère que les taux de rapport de la plantation en pâturage de gleditschias, calculé selon une variété de suppositions concernant la production et les frais, indiquent un bénéfice net variant entre 9% et 24%. On décrit enfin la création du Groupe de Recherche sur le Gleditschia et on discute les recherches qui sont encore à faire.

Introduction

Agroforestry, or tree crop systems, have long been advanced as an answer to the dual problem of soil erosion and declining farm profits (Smith 1950, Felker and Bandurski 1979; Williams and Merwin 1983a; McDaniels and Lieberman 1979; Gold and Hanover 1987). However, few such projects have been implemented. Major obstacles exist to the commercial introduction of agroforestry in developed countries. Field tests have been limited and profitability of new species has not been adequately demonstrated, especially in harvesting and marketing. As with traditional orchards, agroforestry requires considerable time lag between initial cash outlays and financial returns. Implementing agroforestry usually involves allocating land from traditional crops that offer present cash return. Finally, agroforestry requires farmers to adopt non-traditional methods for which there are few successful demonstrations.

Recent developments in agriculture may, however, increase the likelihood of introducing agroforestry projects in developed countries. Some of these factors include soil erosion, falling productivity in agriculture, increasing cost of petro-chemical inputs, and environmental concerns related to the use of chemicals in traditional agriculture (Gold and Hanover 1987). A positive factor facilitating the introduction of pasture tree crops is the improved pasture management techniques utilizing electric fences. These electric fences can provide the necessary protection for inexpensively establishing pasture trees.

[1]Springtree Agroforestry Project, Rt. 2 Box 89, Scottsville, Virginia 24590 U.S.A.

Springtree Agroforestry Project

The Springtree Agroforestry Project has been designed with these factors in mind. Honeylocust (*Gleditsia triacanthos L.*) trees are planted directly into currently operating pastures and hayfields. The trees produce pods which the livestock then harvest themselves under the trees. Such a system offers several advantages. Harvesting and processing costs are nil. Honeylocust pods provide a complementary feed source in the fall when seasonal grass production is declining. Trees are introduced to working farms without interruption of present cash flows. Although not easily quantified, additional benefits from introducing honeylocust to pastures include: reduction of water runoff and topsoil erosion, shade for livestock, a productive pollen and nectar source for bees, a more diversified and aesthetically pleasing pasture environment, and timber upon project termination.

The Springtree Agroforestry Project is being implemented on a 50-hectare farm in central Virginia where sheep and cattle are raised commercially. A comprehensive pasture management program includes an intensive rotational grazing system. This system, now being recommended by many state extension agents and farm journals, replaces the traditional continuous grazing practice where livestock have constant access to large pasture (Murphy *et al.* 1986; and Murphy 1987).

Electric fences subdivide the pastures for intensive grazing and rotation. These fences provide an inexpensive method for introducing tree crops to the pastures by providing protection for young trees from livestock browsing and rubbing. A single or double wire is attached to the fence in a semicircle around each tree and is supported with a fibreglass or metal stake in front of each trunk. High production

A. A. Wilson, "Browse Agroforestry Using Honeylocust," *Forestry Chronicle* 67, no. 3 (June 1991): 232–35. https://doi.org/10.5558/tfc67232-3

honeylocust cultivars have been grafted to seedling nursery stock and planted out. Trees are planted 0.5 to 1 m from the electric fence and 6 to 8 m apart along the fence row, although shading may require some later thinning.

Evaluation of the Browse Agroforestry System

General Characteristics of Honeylocust

Honeylocust is adaptable to a wide range of climate within the temperate zone and is winter hardy to temperatures of below −34°C (Funk 1957). It also thrives in a variety of soils, although it grows poorly on shallow, gravely, or heavy clay soils. Honeylocust it also quite drought-resistant (the natural range includes areas with 51 cm annual precipitation) and can survive severe storms. Its leaves and flowers appear late in the spring and are rarely damaged by late frosts. Although honeylocust normally carries thorns dangerous to livestock and tractor tires, thornless trees can be produced by grafting scions taken from the thornless upper branches of the desired cultivars.

Pod Production

Although incomplete, available data on honeylocust pod production indicate high yields in some cases and considerable variation between southern and northern regions of the temperature zone. Annual yields of 180 kg from mature trees have been reported from South Africa and New Zealand (Loock 1947, Lennard 1980). Ten-year-old trees grown in the southern United States at the Auburn University research project yielded 43 kg per tree (Table 1). Honeylocust have produced considerably lower yields in the shorter growing season of the middle and northern United States (Detwiler 1947).

The yields at Auburn University were achieved with fertilized grafted trees, and some fertilizer applications would undoubtedly be required to obtain the higher yields, especially on marginal lands often recommended for tree crops. A potential problem is the biennial nature of honeylocust pod production (Table 1). This can be offset to some extent by planting different cultivars, and possibly by selective pruning, although the cost effectiveness of such pruning has not been determined (Williams 1980). While male honeylocust are not required for pod formation, they are desirable for full seed development. Selected thornless male cultivars are available, and a ratio of 1 male to 10 to 30 females is recommended. High production cultivars are available, and grafting to seedling nursery stock and transplanting honeylocust is not difficult or expensive.

Harvesting by Livestock

Honeylocust pods are readily harvested by livestock from under the trees. Ripe pods drop gradually throughout the fall, permitting animals to feed on them over several months. In the Auburn University study, honeylocust pod meal as feed for dairy cows was found to be palatable and was substituted for oats on a pound for pound basis (Atkins 1942). Honeylocust pods contain large amounts of sugar and starch, comparable with corn (*Zea mays L.*) (Table 2). Approximately 5% of the honeylocust pod by weight is digestible protein. Hence the digestibility of the honeylocust seed within the pod is critical to the economic viability of the project.

Although cattle eat honeylocust pods, they do not digest the seed (Lennard 1980). For cattle to get the full benefit of the protein value, the pods must first be harvested and then ground into meal. Sheep, however, apparently do digest the honeylocust seeds. Two of the three studies reviewed corroborate our own experience that sheep do digest honeylocust seed when eaten in the pod (leRoux 1959a, Felker and Bandurski 1979, Small 1983a and b). The contradictory findings may be due to differences in test design, in sugar content of pods, and possibly in sheep breeds.

Table 1. Honeylocust production: Auburn University study (Scanlon 1980)[1]

	1942	1943	1944	1945	1946	1947	Average
Tree age	(5)	(6)	(7)	(8)	(9)	(10)	
Calhoun[2]	12.0	0	14.7	28.9	10.0	20.9	14.4
Millwood	26.4	0	66.2	17.9	81.6	5.4	32.9

[1]Average dry kg per tree for 5- to 10-year-old trees.
[2]Cultivars selected for high pod production and sugar content.

Table 2. Estimated nutrient and ethanol yields for corn and honeylocust (Williams and Merwin 1983b).

	Corn	Honeylocust
	$\times 10^3$	
Protein (kg/ha)	0.3	0.4- 2.3
Sugar and starch (kg/ha)	4.5	2.6-15.0
Ethanol (l/ha)	2.8	1.5- 9.0

Grass Production

Evaluation of pasture agroforestry requires determining the effect of trees on grass production. Depending on the tree and grass species selected, as well as the soil, climate, and grazing conditions, this effect may be positive or negative. For example, in a study conducted at the University of Tennessee grass and cattle (weight gain) productivity were greater in pasture under the light shade of black walnuts trees (Zarger and Lutz 1961).

Casual observations of field workers suggest that pasture grasses grow better under honeylocust. The tree's open canopy throws a light shade that may encourage grass production in hot weather. Grasses and legumes grow well right up to the trunk of the tree (leRoux 1959b, Lennard 1980). Honeylocust leaf out late in the spring and leaves die early in the fall, encouraging cool weather grass production. In addition, the tree's small leaflets are easily absorbed into the pasture grasses during leafdrop.

A 17-year study was conducted at Virginia Polytechnic Institute to measure grass production under a honeylocust plantation (Zarger and Lutz 1961). The study found lower grass quality and output (9% on unfertilized plots and 15% on fertilized plots) for pasture under honeylocust. However, because of design problems, the study's conclusions are open to qualification. First, tree spacing in the test plots was too dense and the project was terminated before the effect of proper thinning could be tested. Second, cattle were allowed to graze indiscriminately between the various test plots, and during hot weather they probably remained longer in the shaded plots. The likely result was heavier grazing and greater sod injury and soil compaction in the shaded areas, although none of these factors was measured. If the necessary adjustment for proper tree spacing and eliminating overgrazing are made, honeylocust shade would probably not reduce grass production. For example, at Auburn

University, 6,272 kg/ha of *Lespedeza sericea* was harvested each year from under honeylocust with a density of 85 per ha. or a 11-m spacing (Moore 1948).

Economic Evaluation

The pod production and nutrition data developed above can be used to determine the economic value of honeylocust plantings. Naturally, project costs will be incurred largely during the initial years of the project while financial returns occur only in the later years. Therefore, separate dollar cost and revenue flows must be discounted back to the present to account for the time value of money. A measure of internal rate of return (similar in principle to cost-benefit ratios) is used to compare these dollar flows. The internal rate of return is the highest possible percentage return for a given dollar investment (cost of grafted trees, tree protection, maintenance, and fertilizer) and revenue (measured by the market value of pod production).

Rates of return were calculated using the Auburn University production data and assuming 85 grafted Millwood honeylocust cultivars per ha. To compensate for any possible reduction in grass yield, pod production was estimated conservatively to remain at the level reached after 10 years. Because of their similar nutritional values, the price of oats was used to give a dollar value to the honeylocust pod production. To judge the sensitivity of the internal rates of return to lower production and to lower oat prices, rates of return were also calculated assuming one-half the Auburn University production figures and a 30% reduction in oat price. Because of the uncertainty of fertilizer needs, calculations were made with and without the cost of fertilizer.

Rates of return indicate a net gain from honeylocust plantings for each set of assumptions used. Calculated rates of return range from 25% for fertilized trees under the medium production assumptions to 13% for fertilized trees under the low production assumption (Table 3). Annual pod production of 80 kg. achieved in New Zealand gave a rate of return of over 30%. Clearly, the profitability of livestock grazing improves with the addition of honeylocust to the pasture.

Discussion and Conclusions

Before honeylocust is planted on a large scale, several important research questions remain to be answered.

(1) *Pod Production Data.* Internal rates of return indicate that honeylocust plantings can be economically justified even for pod yields considerably below those demonstrated on 10-year-old trees at Auburn University. However, additional information is needed on pod production, nutrition, and cultivar performance from different sites, especially in the middle and northern United States and in Canada. Toward this end the author has organized the Honeylocust Research Group to systematize data collection and encourage propagation of selected cultivars among professional and amateur researchers.

(2) *Cultivars.* Cultivars with high pod production and sugar content already exist; however, given the considerable genetic diversity found in wild honeylocust, it seems probable that further improvements are possible (McDaniel 1980). One important objective for future selection and breeding is reduction of the tendency for biennial production. Honeylocust provenance plantations leading toward breeding and hybridization work have been established at Michigan State University (Gold 1985).

Table 3. Internal rates of return for pasture honeylocust plantings

	1988 Oat price	Discounted oat price[1]
Auburn University yields assuming fertilizer cost (see Table 1)	25%	20%
One-half Auburn University yields assuming no fertilizer cost	20%	16%
One-half Auburn University yields assuming fertilizer cost	13%	9%

[1]Internal rates of return calculated using 70% of the 1988 oat price to value the honeylocust pod production.

(3) *Cultural Techniques.* More experience with a variety of culture techniques is needed, specifically fertilizer combinations, alternative tree spacing patterns, and pruning as a possible method for reducing the habit of biennial production. Work is also needed to determine the optimal ratio of male to female trees in a pasture.

(4) *Insect Problems.* Insects on honeylocust include the mimosa webworm (*Homadaula anisocentra*), a seed-feeding weevil larvae (*Amblycerus robiniae*), and the European hornet (*Vesta crabro*) (Funk 1957). The effect of these insects on pod production is not currently known. However, as has been the case with other fruit tree species, extensive monoculture plantings may increase insect problems. The mimosa webworm has caused defoliation in some urban honeylocust plantings, and a spraying program of *Bacillus thuringiensis* for this insect may prove profitable for pasture honeylocust.

(5) *Pasture Grasses.* Good pasture grass growth under honeylocust seems likely; however, experimentation with different grass and legume varieties under honeylocust shade is needed.

(6) *Seed Digestibility.* The ability of sheep to digest honeylocust seeds needs further confirmation and any differences for particular breeds noted.

(7) *Farmer Participation.* Many of the traditional objections to agroforestry projects can be met by establishing honeylocust in pastures in use. Farmer participation should be encouraged by low initial costs (using existing fences), continued cash flow from livestock operations, and elimination of harvesting and processing costs. Pasture honeylocust can be established in a traditional or intensive grazing system with little adjustment, allowing farmers to continue normal grazing operations. Honeylocust are now being introduced to working farms in New Zealand and Australia where farm journals carry advertisements from nurseries offering seedling and grafted honeylocust (Smith 1985).

In summary, planting honeylocust along existing pasture fence lines, as done at the Springtree Agroforestry Project, is technically and economically feasible. If adopted, this agroforestry system could help reduce soil erosion by increasing the profitability of livestock grazing as an alternative to traditional row crops.

References

Atkins, O.A. 1942. Yield and sugar content of selected thornless honeylocust. Ala. Agr. Ext. Sta. 53 Ann. Rep.: 25-26.

Davies, D.J. 1982. Forage trees: tree lucerne and other multipurpose leguminous species. *In* D. Noel (Ed.) Tree crops: the 3rd component. Proc. First Australasian Conference on Tree and Nut Crops: 107-120. Subiaco, Australia.

Detwiler, S.B. 1947. Notes on honeylocust. USDA Soil Conserv. Serv. (mimeo).

Felker, P. and R.S. Bandurski. 1979. Uses and potential uses of leguminous trees for minimal energy input agriculture. Econ. Bot. 33: 172-189.

Funk, D.T. 1957. Silvical characteristics of honeylocust. USDA For. Serv., Cent. States Forest Exp. Sta. Misc. Rel. 23.

Gold, M.S. and J.W. Hanover. 1987. Agroforestry systems for the temperate zone. Agroforestry Systems 5: 109-121.

Gold, M.S. 1985. Honeylocust: genetic variation and potential uses as an agroforestry species. Ph.D. Thesis. Michigan State Univ.

Lennard, T. 1980. The story of honeylocust. J. N.Z. Tree Crops Assoc. 5: 14-17.

leRoux, P.L. 1959a. Honey locust tree useful sources of fodder for stock. Farming in S. Afri. 35: 20-21.

leRoux, P.L. 1959b. Advantages and disadvantages of honeylocust. Farming in S. Afri. 35: 8.

Loock, E.E. 1947. Three useful leguminous fodder trees. Farming in S. Afri. 2: 7-12.

MacDaniels, L. and A.S. Lieberman. 1979. Tree crops: a neglected source of food and forage from marginal lands. Biosci. 29: 173-175.

McDaniel, J.C. 1980. A plant breeder looks at some American tree crops: *Morus*, *Gleditsia*, and *Diospyros*. *In:* Tree crops for Energy Co-production on Farms: 113-118. US Solar Energy Res. Inst. Golden, Colorado.

Moore, J.C. 1948. The present outlook for honeylocust in the south. North. Nut Growers Assoc. 39th Ann. Rep. pp. 104-110.

Murphy, B. 1987. Greener Pastures on Your Side of the Fence: Better Farming with Voisin Grazing Management. Arriba Publishing Co., Colchester, Vermont. 232 pp.

Murphy, W.M., J.R. Rice and D.T. Dugdale. 1986. Dairy farm feeding and income effects of using Voisin grazing management of permanent pastures. Am. J. Alt. Agric. I: 147-158.

Scanlon, D.J. 1980. A case study of honeylocust in the Tennessee Valley Region. *In:* Tree Crops for Energy Co-production on Farms: 21-31. US Solar Energy Res. Inst. Golden, Colorado.

Small, M. 1983a. Sheep can digest 90% of the whole honeylocust seeds ingested. Agrofor. Rev. 3: 6.

Small, M. 1983b. Honeylocust pods and the digestion of protein by sheep. Agrofor. Rev. 4: 6-7.

Smith, J.R. 1950. Tree Crops: A Permanent Agriculture. Derin-Adair, New York, NY. 408 pp.

Smith, R. 1985. Contented cow leads to value of honey locusts. Growing Today. 2: 3-4.

Williams, G. 1980. Tree crops for energy production in Appalachia. *In:* Tree Crops for Energy Co-production on Farms: 7-20. US Solar Energy Res. Inst. Golden, Colorado.

Williams, G. and M.L. Merwin. 1983a. Energy- and soil-conserving perennial crops for marginal land in temperate climates. *In* W. Lockeretz (Ed.) Environmentally Sound Agriculture: 317-332. Praeger Co., New York, NY.

Williams, G. and M.L. Merwin. 1983b. Potential tree crops for the production of staple nutrients and alcohol fuel on hill lands. *In:* D.B. Hannaway (Ed.) Foothills for Food and Forests: 365-367. Timber Press. Beaverton, OR.

Zarger, T.G. and A.J. Lutz. 1961. Effects of improved thornless honeylocust shade trees on pasture development. TVA Div. of For. Relat., Tech. Note No. 34. 12 p.

2. Yield and Sugar Content of Selected Thornless Honey Locusts

Yield and Sugar Content of Selected Thornless Honey Locusts. (O. A. Atkins). — For several years studies have been in progress with superior selections of the thornless honey locust. In the spring of 1938, a few trees of the Calhoun and Millwood varieties were planted at Auburn, but sufficient trees were not obtained to complete the planting until the spring of 1940. Since the experiment was started, individual tree records have been kept on yield of pods, height of tree, and trunk diameter. Observations and data indicate that the two varieties included in this experiment produce few pods until the trees are about 4 years old. The data are given in Table 4.

Table 4.—Average Yield per Tree, Trunk Diameter, and Height of Honey Locust Trees, 1942

Variety	Trees in test	Age of trees	Average yield per tree, dry weight basis	Average trunk diameter	Average height
	Number	*Years*	*Pounds*	*Inches*	*Feet*
Calhoun	31	3	1.01	1.13	5.29
Calhoun	13	4	5.20	2.21	8.95
Calhoun	4	5	26.38	3.50	12.33
Millwood	31	3	1.27	1.25	5.58
Millwood	11	4	4.98	2.07	9.11
Millwood	5	5	58.30	3.56	12.38

Two cultural treatments were used in the experiment: (1) clean cultivation, and (2) lespedeza sericea planted for ground cover and cut twice annually, with the material used as a mulch about the trees. No significant difference in growth due to the different treatments was observed. All trees were

O. A. Atkins, "Yield and Sugar Content of Selected Thornless Honey Locusts," Alabama Agricultural Experiment Station.

fertilized annually with a 6-8-4 fertilizer at the rate of 1 pound per tree for each year of the tree's age.

The 5-year-old trees of the Calhoun variety in 1942 produced an average of 26.38 pounds of pods per tree (dry weight basis) which is the equivalent of 1,266 pounds per acre (48 trees per acre); the four highest yielding trees in the experiment averaged 10.5 feet in height and 3.1 inches in trunk diameter; these produced an average of 31.69 pounds of pods or the equivalent of 1,521 pounds of pods per acre.

The 5-year-old trees of the Millwood variety in 1942 produced an average of 58.30 pounds of pods per tree (dry weight basis), which at 48 trees per acre would be the equivalent of 2,798 pounds of pods. The four highest yielding trees averaged 12.5 feet in height and 3.6 inches in trunk diameter, and produced an average of 65.62 pounds of pods or 3,150 pounds per acre.

Preliminary feeding tests with dairy cows for 2 years, using ground honey locust pods as a part of the concentrate mixture, were conducted in cooperation with the Dairy Department. These tests show that the ground honey locust pods may be satisfactorily substituted for oats pound for pound in the concentrate mixture, and that the ground material is very palatable in the ration. Samples of the pods of the Millwood and Calhoun varieties collected at Auburn, Alabama in 1940 and 1941 gave the following average analysis (dry weight basis):

Variety	Invert sugar	Sucrose	Total sugar
	Per cent	*Per cent*	*Per cent*
Millwood	7.45	29.20	36.65
Calhoun	6.40	32.55	38.95

A more complete analysis (average of three composite samples) of honey locust pods, collected at random from high-yielding trees in Alabama, as determined by the Alabama State Chemical Laboratory, gave the following results: moisture, 12.47 per cent; ash, 3.14 per cent; crude protein, 8.58 per cent; crude fat, 2.12 per cent; crude fiber, 17.73 per cent; carbohydrates (not including crude fiber), 55.96 per cent.

The estimated yield of 3,150 pounds per acre which came from trees only 5 years of age had a feed value equivalent to 105 bushels of oats or 56 bushels of corn.

3. Value of Shelter for Wintering Beef Breeding Cows

ANIMAL AND POULTRY HUSBANDRY

Value of Shelter for Wintering Beef Breeding Cows. (J. C. Grimes). — During a 3-year experiment, a group of cows that had access to shelter lost an average of 46 pounds each in the winter, while a similar group that received the same kind and amount of feed but provided no shelter lost 104 pounds each.

14 ALABAMA AGRICULTURAL EXPERIMENT STATION

The rate at which the cows lost weight whether in the open or under the shelter was closely related to the severity of the weather.

J. C. Grimes, "Value of Shelter for Wintering Beef Breeding Cows," Alabama Agricultural Experiment Station.

4. Chapter X: A Summer Pasture Tree for Swine and Poultry—The Mulberry

FIG. 39. *Top*. A systematic orchard of mulberries for hog pasture near Raleigh, N. C. The man stands on top of a mangum terrace.—FIG. 40. *Center*. Hogs in a mulberry orchard planted for them near Fayetteville, N. C.—FIG. 41. *Bottom*. Characteristic burden of fruit produced by the wild American persimmon tree. Fruits nearly an inch in diameter, Augusta, Ga. (Photos J. Russell Smith.)

J. Russell Smith, "Chapter X: A Summer Pasture Tree for Swine and Poultry—The Mulberry," from *Tree Crops: A Permanent Agriculture* (Harcourt, Brace, 1929), 83–93.

CHAPTER X

A SUMMER PASTURE TREE FOR SWINE AND POULTRY—THE MULBERRY

For a large section of the United States the mulberry is easily the king of tree crops when considered from the standpoint of this book; namely, the establishment of new crops which are easily and quickly grown and reasonably certain to produce crops for which there is a secure and steady market for a large and increasing output.

A KING (OF CROPS) WITHOUT A THRONE

The mulberry is excellent food for pigs. To harvest mulberries costs nothing because the pigs gladly pick up the fruit themselves. Therefore, mulberries fit especially well into American farm economics because labor cost is high.

The mulberry tree is no new wildling just in from the woods and strange to the ways of man. It is one of the old cultivated plants. It has resisted centuries of abuse. It has been tried and found to be good and enduring.

It can perhaps be called the potential king of tree crops for the Cotton Belt[1] and part of the Corn Belt. The honey locust, oak, and chestnut probably have greater promise, because their crop can be stored; but the mulberry has already arrived and has proved its adaptability and its worth.

The mulberry is a tree with good varieties already established and waiting to be used.

[1] In actual production today the pecan is far ahead of the mulberry, but the potential market for mulberries is far ahead of that of pecans.

THE ADVANTAGES OF THE MULBERRY TREE

(1) The trees are cheap because they are so easy of propagation.[2]

(2) The tree is very easy to transplant.

(3) It grows rapidly.

(4) It bears as early as any other fruiting tree now grown in the United States, perhaps earliest of all.[3]

(5) The fruit is nutritious and may be harvested without cost.

(6) The tree bears with great regularity as far north as the Middle Atlantic States and New England, also through the Cotton Belt and much of the Corn Belt, and even beyond it into the drier lands.

(7) It has a long fruiting season.

(8) It bears fruit in the shady parts of the tree as well as in the sunshine, and thus has unusual fruiting powers.

(9) It has the unusual power of recovery from frost to the extent of making a partial crop the same season that one crop is destroyed.[4]

(10) The fruit has a ready and stable market since swine and other animals turn it into meat, a product for which there is no prospect of a really glutted market, such as haunts the growers of so many crops.

(11) The trunk of the tree is excellent for posts, and the branches make fair firewood for the farm stove. It is doubtless worth growing in many sections for wood alone.

(12) While attacked to some extent by caterpillars, it prospers at present in most parts of its area without spraying, and seems to have fewer enemies than most other valuable trees.[5]

[2] In 1913 trees of everbearing varieties could be bought for $2.00 per hundred at Green's Nursery, Garner, North Carolina.

[3] They even bear in the nursery row.

[4] This results from the remarkable habit of putting forth secondary buds and producing some fruit after a frost kills the first set of buds.

[5] Unfortunately, according to a letter from the Fruitland Nursery, Augusta, Georgia (1927), the San José and India scale in that locality require

(13) Growth of the mulberry for forage has gone forward in the United States so that for a large area the experimental stage for the *tree* is past (but not for the crop). In localities where the mulberry is not well established experiments are aided by the incomparable boons of low-cost trees, rapid growth, and ease of transplanting.

THE MULBERRY CROP IN THE UNITED STATES

Every claim that I have made for the mulberry has been backed up by correspondence with persons interested in mulberries or with interviews that I have had. In most cases my information comes from the statements of people who grow mulberries or are closely associated with those who did.

Mr. G. Harold Hume of the Glen Saint Mary Nurseries Company, Glen Saint Mary, Florida, wrote April 12, 1913, "All through the Southern States, mulberries are commonly used as feed for pigs and poultry. In North and South Carolina and Georgia nearly every pig lot is planted with these trees, and the mulberries form a very important addition to the pig's diet.[6] There is one variety, Hicks, which will give fruit for about sixty days, in some seasons even for longer.

one good dormant spray per year to keep the tree in good health. This, however, is not a serious burden, especially as it is a winter job.

As the mulberry tree has been cultivated for ages over a wide region, this comparative pest-immunity is probably more dependable than it would be on a tree that has just come in from the forest and has not been subjected to the crowded conditions of artificial plantings. It is also probably a much safer tree than some fresh importation would be.

[6] "In eastern North Carolina it is the common practice to plant orchards of mulberry trees for hogs to run in." (W. F. Massey, Associate Editor, *The Progressive Farmer*, Raleigh, N. C, letter, March 11, 1913.)

"The everbearing mulberry in this country is so common as to occasion very little comment; in fact, they become unpopular on account of their profuse bearing, especially if there are not pigs and chickens enough to pick them up." (Letter, John S. Kerr, Texas Nursery Company, Sherman, Texas, November 19, 1913.)

"The mulberry grows to perfection, fruits abundantly, and is used both for hogs and for poultry." (Professor C. C. Newman, Clemson College, South Carolina.)

With a proper selection of varieties, this season might be extended.

In 1927 Mr. Hume, who is an ex-professor in horticulture, reported that there was little change in the situation.

I find a very general belief in the Cotton Belt that one "everbearing" mulberry tree is enough to support one pig (presumably a spring pig) during the fruiting season of two months or more. Professor J. C. C. Price, Horticulturist, Agricultural and Mechanical College, Mississippi, says,[7] "The everbearing varieties will continue to bear from early May to late July, a period of nearly three months. I believe that a single tree would support two hogs weighing 100 pounds each and keep them in a thrifty condition for the time that they are producing fruit. They could be planted about 35 trees to the acre."

Mr. F. A. Cochran, breeder of Berkshire swine at Derita, North Carolina, said,[8] "I only have a few trees, but they are large ones, 100 feet apart. . . . I would not take $25 per tree for the old trees. I have three hogs to the tree. They are doing fine, in good flesh. . . . I have not weighed any hogs that were fed on mulberries, but estimate that they gained one pound and over per day. My hogs have a feed of two small ears of corn twice a day."

Mr. James C. Moore, farmer of Auburn, Alabama, writes, "I never weighed my pigs at the beginning and close of the mulberry season, but think I can safely say that a pig weighing 100 pounds at the start would weigh 200 pounds at the close. . . . Three-fourths to the mulberries is safe calculation of the gain. I have had the patch about 18 years bearing. I planted my trees just 32 feet apart, and now the branches are meeting, and I have about 40 trees. I have carried 30 head of hogs through from May 1 to August 1, with no food but the gleanings of the barn and what slops came from the kitchen of a small family."

[7] Letter, September 2, 1927.
[8] Letter, July 21, 1913.

Mr. J. C. Calhoun, farmer of Ruston, Louisiana, says, "The variety is the 'Hicks.' I set them out 30x30 feet apart. I have 50 trees. They were mere switches when I set them out about three feet high. They began bearing the second year and made rapid growth. The fourth year after putting them out the trees would nearly touch, and they are abundant bearers—ripening from the last of April to the last of July. There is nothing that a hog seems to enjoy better than mulberries. I always feed my hogs at least once a day, but I find that it takes considerable less feed for them to thrive and do well during mulberry season." (Letter, March 7, 1913.)

In the course of much correspondence and a long journey through the Cotton Belt in 1913 I met many such enthusiastic statements.[8]

In that east-central part of North Carolina where the mulberry orchard is a very common part of farm equipment, a veteran of the Civil War (a captain) declared, "When I lived

[9] "We have talked this matter over thoroughly here in the office and believe that one mulberry tree 10 years old ought to support a pig 4 to 6 months old during the tree's fruit season. As the tree gets older, 15 to 20 years old, if it has had good attention and has grown well, then it ought to support two or three pigs at the same age as mentioned above." (O. J. Howard, J. Van Lindley Nursery Company, Pomona, North Carolina, August 11, 1912.)

"The large mulberry tree of which I spoke is in the south part of Wake County, North Carolina. I stopped at the place to get some water and spoke of the large mulberry trees and made the remark that it was wisely said that one tree fed one pig during fruiting season. The owner said it certainly would, for this tree was feeding fifteen hogs at that time and was probably one hundred years old. The place is near Fuquay Spring, North Carolina." (J. W. Green, Green Nursery Company, Garner, North Carolina.)

It may properly be objected that these men are partisans, being nurserymen with trees to sell, but many farmers without any axes to grind are of the same opinion.

Mr. R. H. Ricks, a farmer specializing in cottonseed at Rocky Mount, North Carolina, gives the following testimony: "I planted two hundred mulberry trees of the everbearing variety thirty-three years ago on what I regarded as waste land. They commenced having some fruit at once, but they did not have profitable crops until the fifth year. I have since planted another orchard of fifty trees. I carry fifty to sixty hogs on the fruit ten to thirteen weeks every year for the last twenty-seven years without other ovah the rivah we had a lot of mulberry trees—300 to 400 mammoth big ones. We had fully 200 hawgs, but we had to send fer the neighbors' hawgs to help out and to keep the mulberries from smellin'."

Where the captain then lived he had a bunch of thirty-five hogs of various sizes running in mulberry and persimmon pasture, all of which I saw. He estimated that one-third of their weight, or one thousand pounds of pork, live weight, was due to the mulberries from eighty trees set twenty-four by thirty-three feet. That runs out about six hundred and twenty-five pounds of pork, live weight, to an acre of rather thin, sandy land with little care and no cultivation. A big yarn, you say? I'll willingly take it back just as soon as any experiment station makes a real test and disproves it.

The trouble is that no station, so far as I can find out, is in a position to disprove it,[10] because none of them has any

food, and in the main bearing season the two hundred and fifty trees would carry twice the number of hogs. Nearly every farmer has a small orchard in this section, eastern Carolina. I regard the mulberry for hog food with much favor."

"I wouldn't take a pretty for my mulberry orchard," said one of these Carolinians. "It's funny to me to see how soon a hawg kin learn that wind blows down the mulberries. Soon as the wind starts up, Mr. Hawg strikes a trot out of the woods fer the mulberry grove. Turn yer pigs into mulberries, and they shed off and slick up nice. It puts 'em in fine shape—conditions 'em like turnin' 'em in on wheat. They eats mulberries and goes down to the branch and cools off, and comes back and eats more—don't need any grain in mulberry time."

"They begin to beah right away," one Carolina farmer declared. "Why, I've seen 'em beah in the nursery row, and they begin to beah some as soon as they get any growth at all."

I have myself seen wild ones bearing in the Virginia woods when only five feet high. As to their hardiness, another Carolina mulberry grower testified as follows: "Yeh can't kill the things. Yeh kin plant yeh mulberries jest like yeh would cane—cut it off in joints and graft from the one yeh want to. I moved a tree last yeah. Jest put a man cuttin' roots off—and he cut 'em scandalous—and I hooked two mules to it and hauled it ovah heah. I didn't 'spect it would live, but it did." ("A Georgia Tree Farmer," J. Russell Smith, *The Country Gentleman,* December 4, 1915, pp. 1921-22.)

[10] The attempt to secure scientifically determined facts from southern stations brought nothing but much favorable opinion and a suggestion that something might have been done at Tuskegee Institute. Dr. G. W. Carver,

facts. This is a reproach to station staffs and also a really interesting piece of human psychology. The mulberry tree warrants careful testing.

THE NEED OF SCIENTIFIC TESTING AND EXPERIMENT

As nearly as I can learn through trusted correspondents in several southern states, there has been little change in the mulberry situation between 1913 and 1927. The neglect of the mulberry as a crop in the face of such evidence seems to require some explanation. However, this psychological and economic phenomenon becomes easier to understand when one recalls the slavish dependence of the southern farmer on the one crop of cotton. By tens of thousands they have resisted the temptations of clover and cowpeas and soy beans and vetch. They still buy hay for the mule. Nor have they planted pecan trees in their door yards. They grow no fruit, and some do not even have anything worth the name of garden.[11] So the mulberry is after all in good company with the things they haven't done. Occasionally one finds a man who has tried mulberries and does not like them because of caterpillars, but in the main I have found enthusiasm among those pioneer farmers who were trying out the crop.

Director of Research and Experiment Station at Tuskegee Institute, Tuskegee, Alabama, wrote (November 2, 1927), "The small amount of analytical data that I have been able to find on the mulberry shows it to be higher in carbohydrates than pumpkins, being fourteen per cent. carbohydrates, a rather convincing evidence that it is really worth while as a fattening food." He also told of their own use of it in growing their own pork supply.

[11] "It is still true that in the principal cotton sections, particularly the black belts, anything that can really be called a garden is quite scarce.

"You are also undoubtedly correct in saying that there are still not only thousands but tens of thousands of cotton farms which make practically no hay and depend almost entirely upon buying hay if any is used, and this in spite of the fact that there has been a tremendous increase in the acreage in alfalfa, soy bean, cowpea, and other hays in the past few years. You could even go further and say that thousands of such farms have no milch cows, practically no poultry and no hogs.

"A real home orchard is yet a comparative scarcity in the Cotton Belt." (Letter from J. A. Evans, Assistant Chief, Office of Coöperative Extension Work, U. S. Department of Agriculture. Sent. 7, 1927.)

THE POSSIBLE RANGE OF THE MULBERRY IN THE UNITED STATES

As with any other little-used crop, the exact range over which the mulberry can eventually spread is today unknown. For example: Kansas seems to have some mulberry territory and some that is not mulberry territory.[12] But there is every reason to expect that breeding can extend this crop.

There is little doubt that at least a million square miles of the United States, and in the most populated parts of the country, are now capable of producing crops of fruit from the everbearing strains of this remarkable tree. Fortunately two of the commoner varieties, the Downing and the New American, originated in New York.[13]

STABLE PRICE AND EASY EXPERIMENT

The price stability of the mulberry should be emphasized in a country where so many commodities find markets that are

[12] State Forester Albert C. Dickens makes the following interesting observations in Kansas Bulletin 165, pp. 324-26. It should be remembered that as a forester he naturally is dealing primarily with wood trees rather than fruit trees, and that therefore his fruit statements might tend to be weak rather than strong.

"The success of the Russian mulberry has been quite varied. In northern Kansas it has been injured very frequently in severe winters.

"In the southern counties of the state Russian mulberry seems much less liable to winter injury. At the fair grounds at Anthony, Kansas, the rate of growth has been especially good, trees set four years ago having attained a height of fourteen feet and a diameter of four inches.

"The fruit is not of high quality, but is often used when other fruits are scarce; and as it ripens with the cherries and raspberries, it seems to attract many birds from the more valuable fruits, and it is frequently planted in the windbreaks about fruit plantations with this end in view. The fruiting season lasts a month or more. The need of some careful selection and breeding of this species is clearly indicated. The species is quite readily grown from cuttings and the better individuals may be propagated and the uncertainty which attends the planting of seedlings be avoided."

[13] "With chickens, ducks, birds, and pigs clamoring for the fruit, the everbearing mulberry is certainly a candidate for experiment in your poultry yard or pig lot; but if you live north of Mason and Dixon's line, make some investigation as to the hardiness of varieties that are offered you." ("A Georgia Tree Farmer," J. Russell Smith, *The Country Gentleman*, December 4, 1915, p. 1822.)

often glutted. Sometimes peaches are not worth the picking. Apples and oranges occasionally rot on the ground, as do beans, peas, and all the truck crops. In contrast to this the mulberry is in the class with corn. We have not had much trouble in the past twenty years about a glut in the corn market, or in that of meat, its great derivative. A stable market is a fact of great importance in considering any crop. The fact that the mulberry has no harvesting costs and needs no special machinery minimizes the risk in experimenting. We need to get the facts on the feeding value of the mulberry from state experiment stations. However, any landowner can try it now. Commercial nurseries have the trees ready.

For the actual use of the mulberry as a farm crop see Chapter XXIII.

The farm yard, the rocky slope, the gullied hill, the sandy waste invite you to try this automatic crop for which there is a world market at a stable price, and probably a very good price when one considers cost of production.

OLD WORLD EXPERIENCE WITH THE MULBERRY

Perhaps some one thinks that I should mention the silk worm. That classic, domesticated insect makes the mulberry leaf worth its hundreds of millions of dollars yearly and thus renders its great service to humanity by enabling hundreds of thousands of hard-worked orientals to eke out a hungry existence. We have the climatic and soil resources for the mulberry trees, but the silk crop is not for us—not in this next hundred years. It requires labor, human labor, lots of it, and in this we have no present prospect of competing with China, Japan, and other very populous countries.[14]

Nor is there much likelihood that in this century we shall put on a tariff that would drive us to such a crop. But if we do ever want to feed silk worms, we have the resources, be-

[14] See Smith, J. Russell, *Industrial and Commercial Geography*, Henry Holt and Company.

cause the humid summer of our Corn and Cotton Belts keeps the trees growing and producing leaves.

The mulberry has another great use among the Asiatics. It is a food of value for a dense population pressing upon resources more heavily than we do. This fruit has long been an important food in many parts of Western Asia.

"Dried white mulberries, practically, but not quite, seedless and extremely palatable, form almost the exclusive food of hundreds of thousands of Afghans for many months of the year. This use of dried mulberries suggests a new tree food crop.[15] Analysis of these dried mulberries (page 93) shows them to have about the food value of dried figs, and the fig is one of the great nutritive fruits. (See table, page 303.)

Ellsworth Huntington of Yale, geographer and explorer, says that in Syria the troubles of the beggar and the dog are over for a time when mulberries are ripe, for both of these mendicants move under the mulberry tree and pick up their living. "Not only do the people eat large quantities of the fruit, but they also dry it and make a flour out of which a sort of sweetmeat is made." [16]

But I am not urging diet reform for people—only for pigs. They are much more amenable to reason, much more easily pressed by necessity. But, nevertheless, this Afghan dried mulberry [17] seems to be a remarkable food according to the ex-

[15] This particular variety, if needed in America, should be expected to thrive in the irrigated lands of our West and Southwest where dry summers and frosty winters somewhat like those of Afghanistan are found.

[16] Personal letter.

[17] "The dried mulberries form the principal food of the poor people of the mountain districts or 'Koistan.' In the valleys of Koistan and around Kabul there are extensive orchards of this mulberry, all irrigated, and the yield seems to be heavy. There is a howl if you have cut down a mulberry tree. When the mulberries are ripe, they sweep under the trees and let the fruit fall down and dry them just as they do the plums in California. For eight months the people live entirely on these mulberries. They grind them and make a flour and mix it with ground almonds. The men come month after month with their shirts filled with them. They can carry in their shirt enough of these dried mulberries for five days' rations. These men are

plorer's record of its use in that country, where it is more important than bread is to the people of the United States.

commandeered and they bring their food with them. They get no other food whatever, mulberries and water are the whole diet. They sit down on the rocks and lunch and dine on nothing but these dried mulberries (Jewett). Here is the analysis of the dried mulberry, thirteen ounces in pulp, from Afghanistan made by F. T. Anderson of this Bureau." (Courtesy of Mr. Peter Bissett.)

Total solids	94.81
Ash	2.75%
Alkalinity of ash as K_2CO_3	.414%
Ether extract	1.60%
Protein (N x 6.25)	2.59%
Acid as malic	.30
Sucrose	1.20%
Invert sugar	70.01%
Starch	Absent
Crude fiber	2.65%

(From records of Division of Seed and Plant Introduction, United States Department of Agriculture. Copy of Inventory Card. 40215 [F. H. B. No. 3445] Morus alba. Mulberry. From Afghanistan. Presented by his Majesty Habibulah Khan, Amir of Afghanistan, Kabul, through Mr. A. C. Jewett. Received February 23, 1915. See Plant Immigrants, 1916, Bureau Plant Industry, U. S. Department of Agriculture, 1916.)

5. Tuskegee Normal and Industial Institute, Experiment Station—Feeding Acorns

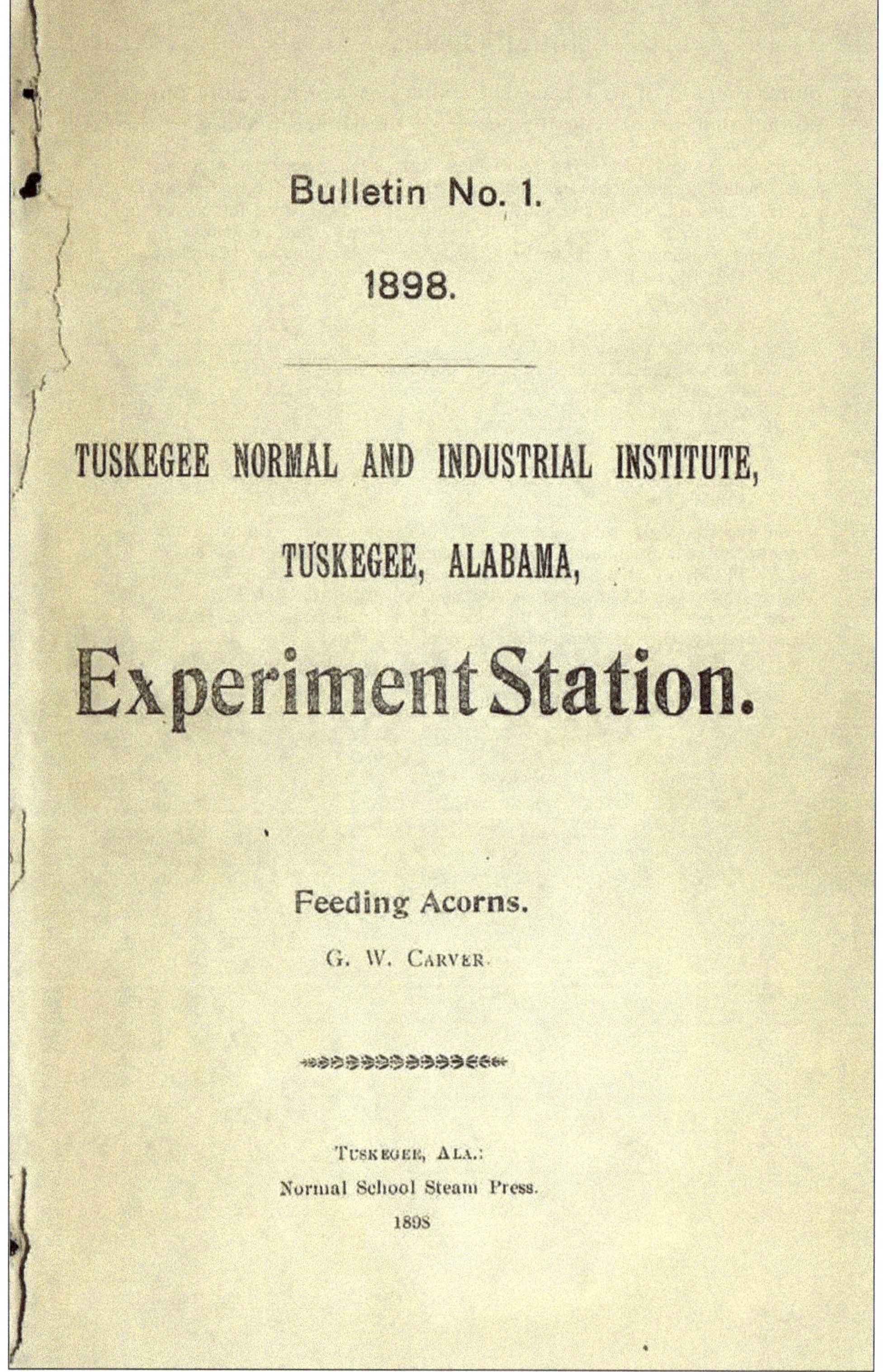
Bulletin No. 1.

1898.

TUSKEGEE NORMAL AND INDUSTRIAL INSTITUTE,

TUSKEGEE, ALABAMA,

Experiment Station.

Feeding Acorns.

G. W. CARVER.

TUSKEGEE, ALA.:
Normal School Steam Press.
1898

George Washington Carver, "Feeding Acorns" (Tuskegee Normal and Industrial Institute, 1898).

Feeding Acorns.

o

G. W. CARVER.

In this beautiful Southland of ours, with so many natural resources, and the repeated failures of the North, East and West to supply the ever-increasing demand for pork, dairy products, etc., has led us to turn much of our attention in this direction.

The great quantity of acorns produced in our oak forests, which have been hitherto practically a waste product, forms the subject of this Bulletin.

I presume that acorns have been used as an article of food ever since their production and that of the animals to eat them, and man has used them in a half-unconscious way, from the time America's first settlers recognized the food value of the wild hog.

It did not take these sturdy colonists long to learn that the meat of the wild hog was fattest and sweetest when the mast (or acorns) were plentiful, as they were then and are now frequently called.

It is not at all strange that so little attention was paid to this food,product, as forests were plenty and the supply far in excess of the demand. The people's custom was to turn their hogs out and let them run at large in the fall, and when wanted for meat, all that was necessary was to take the dogs, wagon and gun; recognize the "split," "half-crop" or some such disfigurement of the ear which served as a mark; shoot them down and return home, to repeat the same process when more were wanted. In some sections of the South this system still exists to some extent.

But as the population increased, the forests gave way to cultivated fields, which, of course, decreased the supply of acorns. Yet our soils were rich and the production of other food stuffs comparatively easy, so that the feeding value of this natural product was, in a great measure, lost sight of, and we think it quite time now to consider more closely their value.

I append here the composition of corn and acorns, in order that the reader may compare them. The figures as given are taken from Dr. Armsby's splendid book on feeding, and the figures are essentially those of Julius Kuhn (Mentzel & v. Lengerke's, Landw. Kalender, 1880.)

Before giving these figures, there are a few terms that need explanation before one can make intelligent comparisons. They are as follows:

Total dry matter.—Refers to the weight of any substance dried in the air.

Protein.—By this we mean the class of substances in feeding stuffs that supplies the material for the growth of tissue.

Fat and Carbohydrats.—Belonging to that class of substances which produce fat and heat in the animal body.

Crude Fibre.—Meaning the woody part, and is more or less indigestible; foods with much woody fibre possesses a low feeding value.

Ash.—This represents the mineral portion, or what is left after the substance has been burned.

This table represents percentages or the number of pounds in 100 pounds of either substance:

	Total dry Matter.	Protein	Fat.	Nitrogen Free Ext.	Crude Fibre	Ash
Corn	88.9	10.5	4.8	70.2	1.9	1.5
Acorns with shell (Fresh)	49.3	2.2	2.0	34.7	9.4	1.0
Acorns (Dry) without shell	85.6	5.6	4.1	69.2	5.1	1.6
DIFFERENCE IN FAVOR OF CORN.						
First	39.6	8.3	2.8	35.5		
Second	3.3	4.9	.7	1.0		

It takes but a glance at the table to show how favorably acorns compare with corn as a fattening food; yet millions of bushels of this choice food product goes to waste in the United States every year.

The school possesses about 400 head of hogs, which are being grown, fattened and prepared for the table, solely on acorns and kitchen slops.

We purchased about 1,000 bushels this fall, and would have doubled the amount had we not feared their keeping qualities when put in bulk, but experience proves that it is only necessary to keep them cool and dry.

Another interesting experiment is the feeding of acorns to milch cows as a grain ration; in other words, instead of giving them corn, and such large quantities of cotton seed and cotton seed meal, acorns are given.

It is impossible at this writing, to give an accurate, well balanced ration, as the exact feeding value of acorns has, as yet, not been determined. We are feeding the following amount with good results: Two quarts at a feed twice per day, making five pounds.

The results have been very gratifying. While the milk did not materially increase as to quantity, it greatly improved in the amount of butter fat, and as far as I am able to determine at this writing, no bad effects on the butter is produced.

This cow has been fed as above described for one month and sixteen days; the amount of fat gained is also quite noticeable, which leads me to believe that beef may be very profitably produced on this natural food product.

Prof. Radusch, late of Harvard University, has quite an extensive poultry ranch on Greycloud Island, Minnesota, and in an interview with him he makes, substantially, the following statement:

I have tried nearly or quite all of the various poultry foods advertised, and find that acorns, ground up, produce more eggs than anything I have tried; besides keeping the flock healthy.

It is safe to conclude that if fed in too great quantities to laying hens, too much fat would be produced.

The above statement is not at all surprising, if we call certain facts to mind. Many who will read this Bulletin, remember the days of the wild pigeon, when they flew in such great droves as to darken the sun, and remind one of a great moving cloud, indefinite in length, depth and breadth; how they would settle so thickly in their favorite roosting places as to frequently break large limbs several inches in diameter, from the trees upon which they perched.

At this time of the year, which was in the fall, many depended almost wholly on these birds for their supply of meat, and a most excellent dependence it was, as they were easily gotten and the meat of high quality, being so fat that no additional oil was needed in their preparation for the table. Everyone hailed with delight a big crop of acorns, for this meant a great many fat pigeons, and that they would remain a long time.

In feeding acorns there is this precaution necessary; where large quantities are given, plenty of laxative food should be included in the ration, as they are rather binding in their nature and likely to produce harmful results.

Considering the low price of cotton, the high price of corn, the feeding value of acorns and the great quantity produced, the entire South should bend much of their efforts toward saving this valuable natural food. They should be gathered in the fall, put in well ventilated bins, barrels or boxes, and kept as cool as possible to retard the destructive work of the worms and to prevent sprouting. The small varieties are more desirable here than the larger sorts, as they are less liable to sprout in bulk, and are more resistant to worms.

In feeding them to hogs, we find that rather a soft, spongy flesh is produced, with an oily-like lard, that hardens with great difficulty, and frequently not at all. This is readily overcome by feeding corn two or three weeks before butchering, although many hundred pounds of meat go into market, without complaint, that has never been topped off with corn.

A number of persons in this vicinity are feeding them to both cattle and hogs, and in every case report good results.

6. Conservation Practice Fact Sheet: Silvopasture — 381: Establishing a Silvopasture System

United States
Department of
Agriculture

Silvopasture - 381

Establishing a Silvopasture System

Conservation Practice Fact Sheet ***December 2019***

INTRODUCTION

Silvopasture is a management option by which landowners can integrate woody plants, forages, and grazing livestock as a system on the same land unit. Typically, silvopasture systems are established on existing timber plantations (usually pine plantations) by thinning the trees to a desirable level to support an understory of forage, or on existing pastures by planting strips of trees and maintaining forage corridors between them. Silvopasture may also be implemented to add forage in wooded areas that are already used by livestock, typically for shade, *but the practice should not be used in Maryland to convert natural forests to grazing lands.*

Prescribed grazing is a key management practice that must be implemented with silvopasture to minimize compaction damage to tree roots and improve forage production. A properly managed silvopasture system does not allow livestock unmanaged access to woods, nor does it involve using livestock to control invasive species. Livestock in a silvopasture system must be managed with rotational grazing that will maintain or improve the quality of the soil and plant communities.

WHY IMPLEMENT SILVOPASTURE?

Depending on a landowner's objectives, and how the system is designed and managed, silvopasture can provide a variety of benefits. A silvopasture system can be implemented to:

- **Improve livestock health and productivity** by reducing stress from summer heat and cold winter temperatures. Trees in a silvopasture system can provide shade as well as wind protection for many types of livestock, including beef and dairy cattle, goats, and pigs.

- **Improve forage quality.** Cool-season forages that are grown in partial shade can have equal or better quality and digestibility when compared to forages grown in full sun. In addition, cool-season forages in partly shaded areas may continue to grow for a somewhat longer period into early summer, thus extending the grazing season and reducing the need for supplemental forage. There are also opportunities for planting edible trees and shrubs that can be used with traditional forages to create a more varied diet for livestock.

- **Improve soil health, reduce erosion, and improve water quality.** In traditionally grazed pastures with little shade, livestock tend to congregate in the few shady locations, resulting in poor forage conditions in those areas and the potential for soil compaction, erosion, and water quality problems. A silvopasture system can be designed to spread trees throughout a pasture, allowing more evenly dispersed grazing and improved forage management.

- **Improve wildlife and pollinator habitat.** Planting a wide variety of native trees and shrubs in an existing pasture can provide food and cover for many wildlife and pollinator species, such as songbirds, turkeys, quail, rabbits, squirrels, butterflies, and bees. In addition, landowners who are interested in hunting, or in leasing their land for hunting, may realize improved hunting opportunities.

- **Increase income potential.** Trees can be planted for the sole purpose of providing shade for livestock, but the silvopasture concept is based on trees also being managed for wood products, such as ornamental greenery, pine straw, posts and poles, veneer logs, or high-quality sawtimber. The extent to which this is feasible depends on a landowner's ability to manage the trees and harvest the materials, as well as the availability of local markets for selling the products. Depending on the future use of the trees (e.g., for poles or sawtimber), management activities such pruning and stand thinning may be needed in future years to improve the quality of wood products, as well as to maintain adequate light to the forage understory.

Program Participation – *If you are enrolled in a program that provides financial assistance for your silvopasture, specific restrictions and requirements may apply. Refer to the program guidance provided in addition to this fact sheet.*

Natural Resources Conservation Service (NRCS) – Maryland

USDA NRCS, "Conservation Practice Fact Sheet: Silvopasture — 381: Establishing a Silvopasture System," December 2019. https://efotg.sc.egov.usda.gov/api/CPSFile/15213/381_MD_GD_Silvopasture_Establishment_Fact_Sheet_2019_pdf

Silvopasture Fact Sheet - 2

Note: Commercial production of tree fruits and nuts (e.g., apples, peaches, sweet cherries, walnuts, etc.) for human consumption is not recommended on silvopasture because of food safety concerns. Livestock manure may contain harmful pathogens that can contaminate fruits/nuts and result in food-borne illness, especially if products are eaten raw. Although the risk of contamination of tree-grown fruits and nuts is lower than for produce grown on or near the ground, contamination can occur via wind-borne dust from manure, packing containers placed on the ground, or on dropped fruit. Food safety regulations (and some buyers' specifications) may restrict livestock access into areas where fruit and nut crops are grown, especially during the growing season for these products.

SYSTEM DESIGN

Overview

Establish and maintain silvopasture in a semi-forested condition that is at least 10 percent stocked with single-stemmed woody species that will be at least 4 meters (13 feet) tall at maturity. Shrubs may also be included in a silvopasture system, provided that the minimum requirement for trees is also met. An appropriate stocking rate can be based on the number of trees of a specific size per acre, or on total basal area per acre (i.e., the cross-sectional area of trees measured at 4.5 feet above the ground, and expressed in square feet per acre).

Locate facilities for providing water, minerals, or supplemental feed such that livestock will properly utilize forages in the silvopasture. Control livestock access to areas with sensitive soils (e.g., wetlands, riparian zones, habitats of concern, karst areas, etc.). Refer to the NRCS Conservation Practice Standard for Prescribed Grazing (528) to develop and implement a plan to manage grazing at appropriate levels to establish and maintain silvopasture productivity and function.

The woody component of a silvopasture can be laid out in rows, or irregularly scattered throughout a field, or in some other configuration as needed, based on factors such as site conditions (e.g., topography; locations of existing trees, fence lines, water sources, sensitive areas, etc.), types of livestock, and landowner preferences. Design the tree component to optimize the amount of sunlight reaching the ground to maintain desired forages, while providing the desired shelter and/or shade for livestock. Row-type systems that are laid out in an east-west orientation will generally allow maximum sunlight for forage production. To the extent feasible on sloping land, lay out tree rows and forage alleys along the contour to reduce erosion.

Two commonly used silvopasture layouts are as follows:

- **Row Layout** – Utilizes one or more rows of trees (may include shrubs) in strips along the contour of a field, alternating with alleys of forage. Trees are usually established in single, double, or triple row plantings (sometimes referred to as "sets"), with rows spaced at least 10 feet apart. Forage alleys are typically 30 to 80 feet wide to accommodate management with agricultural equipment and to allow adequate sunlight for forages.

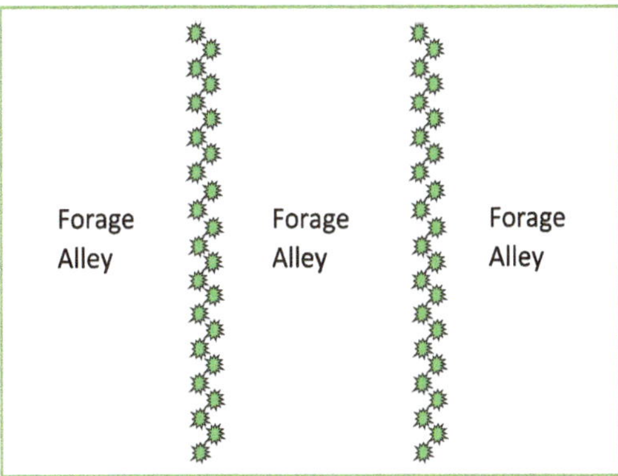

Figures 1a and 1b. Example row layouts with single and double row plantings. Trees that are initially planted closely together can be thinned as the stand ages.

NRCS, MD
December 2019

- **Irregular Layout** – Trees (may include shrubs) are widely spaced as single plants or in clumps throughout the pasture. This layout can provide a more natural-appearing landscape with woody plants and grasses interspersed throughout the pasture. However, management with agricultural equipment and exclusion of livestock from the trees can be more time-consuming and costly if trees are scattered and not in straight rows.

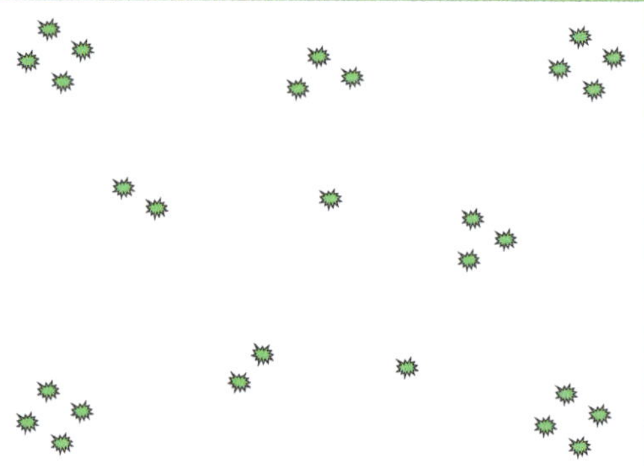

Figures 2a and 2b. Example irregular layouts with single trees and clumps scattered throughout the pasture.

Woody Plant Component

Site Preparation. Where trees, or a combination of trees and shrubs, will be added to existing pasture or cropland, perform site preparation and tree/shrub planting as needed based on existing vegetation and soil conditions. In existing pasture, grass will be very competitive with woody seedlings. Therefore, an herbicide-treated strip 4 to 6 feet wide per tree row (or per tree for widely-spaced single plantings), is strongly recommended before planting.

Alternatively, cultivation may be used to prepare a planting strip, but it is usually less effective than herbicides for killing heavy sod or persistent weeds. Consider planting a low-growing, non-competitive companion grass such as red fescue in the woody planting strip to suppress weeds, especially when cultivation is used. While it is preferable to have minimal vegetation immediately adjacent to woody seedlings, bare ground in the planting strip can provide a good seedbed for germination of noxious weeds and invasive plants (e.g., various thistles, multiflora rose) that may be present in pasture soil.

For existing wooded areas that will be converted to silvopasture, thin and/or prune existing trees to reduce canopy cover sufficiently to allow forage establishment and adequate growth. Thinning to a basal area of 30 to 50 square feet per acre is typically suggested, although this can vary based on the landowner's objectives and a professional forester's recommendation. Wooded areas often exist on wet or sloping sites; do not convert these areas to silvopasture, because adding or intensifying grazing there will result in soil degradation. Refer to the Forage Component section of this fact sheet for further discussion of site preparation for forage planting.

Types of Plant Materials and Estimating Quantities. Planting stock may include bare-root seedlings, tublings and other container plants, and balled-and-burlapped stock. Seedlings and other small stock have the advantage of being easier to plant than larger container and balled-and-burlapped plants but take longer to reach a useful size. Depending on the purpose of the planting, it may be preferable to use larger stock planted widely apart vs. seedlings planted closely together.

Planting rates are usually based on the establishment goal (number of living trees/shrubs desired at two years after planting), the type of planting stock, and its expected survival rate. Determine the planting rate by dividing the establishment goal by the expected survival rate for the type of planting stock to be used. The survival rate of bare-root seedlings is generally estimated at 65%, and at 95% for container and balled-and-burlapped stock. For example, to meet an establishment goal of 50 trees per acre, 77 trees should be planted if bare-root seedlings will be used and the expected survival rate is 65% (i.e., 50 / 0.65 = 77). It may be necessary to adjust the planting rate if survival is expected to be significantly different than the 65% or 95% rates.

NRCS, MD
December 2019

Silvopasture Fact Sheet - 4

If long-term timber production is desired, consider existing markets for intermediate products when determining the planting rate. Keep in mind that higher planting densities will require periodic thinning of smaller-diameter trees. If there are local markets for thinned wood, then the higher planting densities may be economically advantageous. However, if markets for intermediate products aren't available, then planting at a lower density will reduce the need for non-commercial thinning.

Species Selection and Spacing. Select woody species and determine the appropriate spacing based on the intended purpose of the planting. Table 1 (below) provides some guidelines. Be aware that the foliage, bark, and seeds of many woody plants can be toxic to livestock. Livestock do not typically eat toxic plants but are more likely to do so when more desirable forage is not available. Plants that are frequently listed for their toxicity include cherries (e.g., black cherry, chokecherry, and other *Prunus* spp.); they are highly toxic to all ruminants, horses, and swine, and should not be planted in or adjacent to pastures. Oaks can be moderately toxic, primarily to cattle, sheep, and horses, especially if spring foliage is consumed in large quantities. Red maple, box-elder, black locust, and black walnut can also be toxic to horses.

Table 1. Recommended trees species and spacing, based on the purpose of the planting.

Purpose of the Planting	Recommended Species	Spacing
Shade for Livestock	Trees that have moderate to fast growth rates and develop broad canopies are ideal for shade. Poplar hybrids and willow hybrids are among the fastest-growing. Many deciduous trees (e.g., most oaks, hickories, maples) can be expected to have a canopy spread of 15-20 feet in diameter at 20 years, and 40-60 feet at maturity, although growth will vary based on the species and site conditions.	Shade trees can be planted widely apart, based on expected canopy spread when the tree reaches a useful size. Or, they can be planted closely together and then thinned periodically, typically at 10-year intervals, to allow canopy spread and to maintain adequate sunlight for forages. A tree-to-tree spacing of 20 feet or more can allow 20 years of canopy growth for most deciduous species without the need for early thinning.
Shelter for Livestock	Evergreens are especially useful for providing protection from winter winds. Choose species that have moderate to fast growth rates and medium to high density in winter. Arborvitae, eastern red cedar, Atlantic white cedar, Norway spruce, and white spruce are generally recommended.	Plant a double row of evergreens with an in-row spacing of 10-15 feet (and 10-20 feet between rows) for wind protection.
Wildlife Habitat	Select a diverse mix of native species that have medium to high value for wildlife food and cover, including pollinator food sources. Various hickories, oaks, and beech can provide nuts for wildlife food; dogwoods and service-berries produce fruits. Evergreens can provide year-round cover. Early spring pollen sources for pollinators include redbud, persimmon, sassafras, and tulip poplar.	Wildlife plantings can be spread out (e.g., 20 feet apart or more) to encourage canopy growth and early production of seeds/fruits.
Hardwood Timber Products	Species with high value for hardwood timber include various hickories, oaks, maples, and black walnut.	Seedlings are usually planted closely together (e.g., on 8 or 10-foot grid spacings) to encourage straight trunks with minimal branching. These high-density plantings require periodic thinning to remove excess trees and reduce competition, and to maintain adequate sunlight for forages.
Fodder for Livestock	Species with high value as livestock fodder include various willows (shrub and tree types), black locust, and honeylocust.	Trees planted for fodder can be widely spaced (e.g., 20 feet apart) to allow canopy spread and early production of fodder.

Protective Measures. Protect trees to prevent damage from trampling or browsing by livestock and wildlife, especially during establishment. Protective measures can include fencing, tree shelters, and/or removal of livestock from pastures until the average height of the tree's terminal bud exceeds the browsing height of the livestock or is of sufficient size to resist breakage when livestock rub on it. Protection from livestock may be needed for 10 to 20 years, depending on the original size of the planting stock, the growth rate of the trees, and the species of livestock. If trees/shrubs are used to feed livestock, provide protection for an adequate regrowth period after grazing, pollarding, or coppicing them.

Refer to the *Maryland Conservation Planting Guide* for information concerning tree and shrub species selection, growth rates, canopy density, wildlife habitat benefits, planting dates, rates, methods, and care in handling and planting of planting stock. Refer to the Maryland NRCS Fact Sheets *Trees and Shrubs – Establishing and Maintaining Bare-root Seedlings* and *Trees and Shrubs – Establishing and Maintaining Containerized and Balled-and-Burlapped Plantings* for establishment and maintenance recommendations.

Forage Component

Implement forage planting when forage is not present in the desired quantity or quality (e.g., as on overgrazed pasture, or on cropland or a grazed wooded area that will be converted to silvopasture). If trees/shrubs will also be planted, consider the amount of site disturbance that can occur when planting them. If mechanical tree-planting equipment will be used, it may be preferable to plant the trees/shrubs first to avoid tearing up newly planted forages.

Site Preparation. Determine the need for site preparation based on existing vegetation and soil conditions. Seed-to-soil contact is critical for any planting. On existing pastures, stand renovation may require herbicide treatment to kill existing vegetation and/or tillage to prepare a seedbed prior to planting.

On grazed wooded areas, it may be impossible to prepare a seedbed without destroying the roots of existing trees. In this situation, the best strategy may be raking to remove existing debris, then broadcast seeding and using a cultipacker or small tractor to press the seed into the soil. Especially on slopes, lightly mulch with straw or hay to minimize seed displacement by rain. Hydroseeding may also be an option if the site is large enough and equipment is available. Forage planting may not be feasible on shallow, rocky soils, or on poorly drained soils that are not currently supporting pasture grasses; livestock should not be allowed on such sites.

Shade Tolerance and Forage Species. Cool-season forages are generally more shade tolerant than warm-season forages – with the exception of eastern gamagrass. To support forage productivity in the shade beneath the trees, canopy cover should be no more than approximately 35% for warm-season grasses and 50% for cool-season grasses and forbs. Native warm-season grasses (e.g., switchgrass, big bluestem, indiangrass) or introduced cool-season grasses and legumes (e.g., Kentucky bluegrass, tall fescue, orchardgrass, perennial ryegrass, red clover, white clover) may be suitable for a silvopasture system, provided the shading guidelines are not exceeded. Cool-season forages are generally less competitive for moisture in summer months than warm-season grasses.

Forages grown in part shade will have less exposure to sunlight than plants in an open pasture and may need longer periods for regrowth. On established forage stands, a short, intensive grazing period followed by a longer rest period can help maintain forage productivity.

Livestock Exclusion. Livestock should be excluded from grazing new forage plantings for at least one year. Forage may be mechanically harvested while livestock are excluded during the establishment period, provided there is enough space between rows to accommodate harvesting equipment. As noted above, livestock exclusion will also be needed on a regular basis to allow regrowth of established forage.

Table 2. Relative tolerance to moderate shade for selected forage species (ranked high to low).

Forage Species	Shade Tolerance (45-55% shade)
Kura clover	High
Crimson clover	Δ
Kentucky bluegrass	▪
Eastern gamagrass	▪
Subterranean clover	▪
Red clover	▪
Tall fescue	▪
Orchardgrass	▪
Smooth bromegrass	▪
Berseem clover	▪
Timothy	▪
Alfalfa	▪
Buckwheat	▪
Birdsfoot trefoil	▪
Perennial ryegrass	▪
Annual ryegrass	▪
Alsike clover	▪
Creeping red fescue	▪
Big bluestem	▪
White clover	▪
Little bluestem	▪
Bermudagrass	▪
Indiangrass	▪
Korean lespedeza	∇
Switchgrass	Low

Silvopasture Fact Sheet - 6

Refer to the *Maryland Conservation Planting Guide* for requirements concerning forage species selection, planting dates, rates, methods, and care in handling and planting of seed or planting stock. Refer to the following Maryland NRCS Fact Sheets: *512 - Forage and Biomass Plantings, Cool-Season Grasses,* and *Warm-Season Grasses* for establishment and maintenance recommendations.

Use the *Maryland Silvopasture Implementation Requirements* sheet to prepare and document the planting plan for the Silvopasture practice.

SELECTED REFERENCES

Arbor Day Foundation. *Tree Database* (website). https://www.arborday.org/trees/treeGuide/browsetrees.cfm

Clason, T.R., and J.L. Robinson. 2000. *From a Pasture to a Silvopasture System.* USDA – National Agroforestry Center. Agroforestry Note 22. Available at https://digitalcommons.unl.edu/cgi/viewcontent.cgi?article=1020&context=agroforestnotes

Clason, T.R., and J.L. Robinson. 2000. *From a Pine Forest to a Silvopasture System.* USDA – National Agroforestry Center. Agroforestry Note 18. Available at https://digitalcommons.unl.edu/cgi/viewcontent.cgi?referer=https://www.google.com/&httpsredir=1&article=1016&context=agroforestnotes

Cornell University. *Plants Poisonous to Livestock* (website).https://poisonousplants.ansci.cornell.edu/index.html

Federal Register. November 27, 2015. 21 CFR Parts 11, 16, and 112, *Standards for the Growing, Harvesting, Packing, and Holding of Produce for Human Consumption; Final Rule.*

Fike, J.H.,A.L. Buergler, J.A. Burger, and R.L. Kallenbach. 2004. *Considerations for Establishing and Managing Silvopastures.* Plant Management Network. 1-12. Available at http://www.ext.vt.edu/topics/agriculture/silvopasture/files/silvopastures-considerations.pdf

Gabriel, Steve. 2018. *Silvopasture: A Guide to Managing Grazing Animals, Forage Crops, and Trees in a Temperate Farm Ecosystem.* 320 pages.

Oregon State University, Extension Service. October, 2009. *Silvopasture: An Agroforestry Practice.* https://catalog.extension.oregonstate.edu/sites/catalog/files/project/pdf/em8989.pdf

Penn State Extension. March, 2018. *Understanding FSMA: The Produce Safety Rule.* https://extension.psu.edu/understanding-fsma-the-produce-safety-rule

USDA National Agroforestry Center. 2008. *Working Trees: Silvopasture, An Agroforestry Practice.* Available at https://digitalcommons.unl.edu/cgi/viewcontent.cgi?referer=https://www.google.com/&httpsredir=1&article=1009&context=workingtrees

U.S. Department of Health and Human Services, Food and Drug Administration, Center for Food Safety and Applied Nutrition (CFSAN). October, 1998. *Guidance for Industry: Guide to Minimize Microbial Food Safety Hazards for Fresh Fruits and Vegetables.* https://www.fda.gov/regulatory-information/search-fda-guidance-documents/guidance-industry-guide-minimize-microbial-food-safety-hazards-fresh-fruits-and-vegetables

Van Sambeek, J. W., N. E. Navarrete-Tindall, H. E. Garrett, C.-H. Lin, R L. McGraw, and D. C. Wallace. December 2007. *Ranking the Shade Tolerance of Forty-five Candidate Groundcovers for Agroforestry Plantings.* The Temperate Agroforester, Vol. 15, No. 4. https://www.nrs.fs.fed.us/pubs/9201

Virginia Nursery and Landscape Association. *Tree Canopy Spread and Coverage in Urban Landscapes.* http://dendro.cnre.vt.edu/predictions/canopy.cfm

NRCS, MD
December 2019

7. Leaf Nutrient Analysis

Date	Species	% Crude Protein	%ADF	%NDF	Relative Feed Value
2019	Black Locust	23.61	18.70	32.89	227.75
2020		24.30	21.98	30.50	222.73
2019	Buckthorn	18.35	14.58	41.05	272.13
2020		22.80	22.47	28.02	247.45
2019	Honeysuckle	12.71	20.04	39.99	196.63
2020		15.43	26.63	33.53	198.36
2019	Poplar	14.99	22.55	35.73	182.38
2020		15.90	29.78	36.13	176.36
2019	Wild Cherry	13.84	17.49	39.56	211.50
2020		14.94	28.67	48.52	133.82
2019	Willow	15.63	21.49	37.78	200.25
2020		17.15	32.17	38.34	172.80
2000 - 2019 NY samples	Legume Pasture	26.23	27.75	35.96	192.59
2000 - 2019 NY samples	Grass Pasture	15.79	35.65	60.89	98.72

Recommended NDF and Crude Protein values for various ruminant life stages		
	NDF	**Crude Protein**
General	under 70%	more than 8%
Reproduction	under 50%	10 - 12%
Growth	30 - 40 %	16 - 18%
Lactation	under 55%	12 - 14%

Macro nutrients

Date	Species	Calcium	Phosphorus	Magnesium	Potassium
2019	Black Locust	1.29	0.23	0.15	1.43
2020		1.52	0.25	0.18	1.64
2019	Buckthorn	2.92	0.21	0.29	2.46
2020		2.41	0.24	0.29	2.71
2019	Honeysuckle	2.50	0.18	0.33	1.69
2020		2.04	0.22	0.32	1.87
2019	Poplar	1.71	0.20	0.22	1.52
2020		1.74	0.21	0.23	1.26
2019	Wild Cherry	2.17	0.19	0.24	1.25
2020		1.77	0.22	0.30	1.19
2019	Willow	2.13	0.23	0.34	1.14
2020		1.67	0.24	0.30	1.41
2000 - 2019 NY samples	Legume Pasture	1.21	0.38	0.26	1.79
2000 - 2019 NY samples	Grass Pasture	0.5	0.31	0.14	0.93

Micro nutrients

Date	Species	Iron	Zinc	Copper	Manganese	Molybenum
2019	Black Locust	86.00	33.88	9.88	93.88	0.64
2020		76.67	28.17	7.67	70.83	0.48
2019	Buckthorn	153.63	18.88	7.25	78.13	2.03
2020		114.42	20.17	6.92	83.75	0.55
2019	Honeysuckle	184.88	22.88	7.50	70.25	12.08
2020		98.75	19.92	7.67	61.50	0.78
2019	Poplar	70.00	135.63	8.38	98.63	0.80
2020		55.17	99.08	6.42	77.50	0.49
2019	Wild Cherry	100.50	21.75	8.38	226.38	0.78
2020		81.92	19.00	7.08	131.50	0.53
2019	Willow	93.88	180.25	6.50	155.88	0.79
2020		80.27	147.91	7.82	201.45	0.43
2000 - 2019 NY samples	Legume Pasture	206.86	50.94	5.89	18.03	2.09
2000 - 2019 NY samples	Grass Pasture	0	0	4.76	7.76	0

8. Honey Locust Analysis

280 Newport Rd, Leola, PA 17540
Main 717.656.9326 ° Fax 717.656.0910
www.waypointanalytical.com

Project Information : Austin Unruh

Original Report Date : 12/30/2020
Revised Report Date: 12/30/2020
Received : 12/04/2020

Report Number : **20-339-0231**

REPORT OF ANALYSIS

Lab No : **91229**
Sample ID : **Hershey**

Matrix: **Solids**
Sampled:

Test	Results	Units	MQL	DF	Date / Time Analyzed	By	Analytical Method
Moisture	**43.4**	%		1			AOAC 930.15
Dry Matter	**56.6**	%	0.10	1	12/07/20 15:19	KNP	AOAC 930.15
Crude Protein	**9.61**	% - dry	0.17	1	12/08/20 13:30	KNP	AOAC 990.03
Crude Fat	**1.30**	% - dry		1	12/21/20 13:00	RLF	AOAC 920.39
ADF Fiber	**27.4**	% - dry	0.176	1	12/08/20 10:22	RLF	AOAC 973.18
Neutral Detergent Fiber	**30.2**	% - dry	0.176	1	12/07/20 11:47	RLF	NDF
TDN	**68.4**	% - dry	0.176	1	12/08/20 10:22	RLF	AOAC 973.18
Estimated Net Energy (ENE)	**67.0**	ton/cwt - dry		1	12/08/20 10:22	RLF	AOAC 973.18
Net Energy Lactation (NEL)	**0.704**	Mcal/lb - dry		1	12/08/20 10:22	RLF	AOAC 973.18
Net Energy Maintenance (NEM)	**0.715**	Mcal/lb - dry		1	12/08/20 10:22		AOAC 973.18
Net Energy Gain (NEG)	**0.443**	Mcal/lb - dry		1	12/08/20 10:22		AOAC 973.18
Calcium	**0.44**	% - dry		1	12/08/20 11:30	KNP	FEED METALS
Phosphorus	**0.15**	% - dry		1	12/08/20 11:30	KNP	FEED METALS
Potassium	**1.36**	% - dry		1	12/08/20 11:30	KNP	FEED METALS
Magnesium	**0.07**	% - dry		1	12/08/20 11:30	KNP	FEED METALS
Sugar	**37.5**	% - dry		1			AOAC 950.01

Qualifiers/ Definitions

DF	Dilution Factor	L	Limit Exceeded
MQL	Method Quantitation Limit		

Page 1 of 3

280 Newport Rd, Leola, PA 17540
Main 717.656.9326 ° Fax 717.656.0910
www.waypointanalytical.com

Project Information : Austin Unruh

Original Report Date : 12/30/2020
Revised Report Date: 12/30/2020
Received : 12/04/2020

Report Number : **20-339-0231**

REPORT OF ANALYSIS

Lab No : **91230**
Sample ID : **Ash**

Matrix: **Solids**
Sampled:

Test	Results	Units	MQL	DF	Date / Time Analyzed	By	Analytical Method
Moisture	**36.7**	%		1			AOAC 930.15
Dry Matter	**63.3**	%	0.10	1	12/07/20 15:19	KNP	AOAC 930.15
Crude Protein	**11.8**	% - dry	0.15	1	12/08/20 13:33	KNP	AOAC 990.03
Crude Fat	**2.05**	% - dry		1	12/21/20 13:00	RLF	AOAC 920.39
ADF Fiber	**28.6**	% - dry	0.157	1	12/08/20 10:22	RLF	AOAC 973.18
Neutral Detergent Fiber	**43.4**	% - dry	0.157	1	12/07/20 11:47	RLF	NDF
TDN	**67.6**	% - dry	0.157	1	12/08/20 10:22	RLF	AOAC 973.18
Estimated Net Energy (ENE)	**65.9**	ton/cwt - dry		1	12/08/20 10:22	RLF	AOAC 973.18
Net Energy Lactation (NEL)	**0.696**	Mcal/lb - dry		1	12/08/20 10:22	RLF	AOAC 973.18
Net Energy Maintenance (NEM)	**0.706**	Mcal/lb - dry		1	12/08/20 10:22		AOAC 973.18
Net Energy Gain (NEG)	**0.434**	Mcal/lb - dry		1	12/08/20 10:22		AOAC 973.18
Calcium	**0.50**	% - dry		1	12/08/20 11:30	KNP	FEED METALS
Phosphorus	**0.22**	% - dry		1	12/08/20 11:30	KNP	FEED METALS
Potassium	**1.40**	% - dry		1	12/08/20 11:30	KNP	FEED METALS
Magnesium	**0.11**	% - dry		1	12/08/20 11:30	KNP	FEED METALS
Sugar	**28.8**	% - dry		1	12/30/20 12:35	GLZ	AOAC 950.01

Qualifiers/ Definitions

DF	Dilution Factor	L	Limit Exceeded
MQL	Method Quantitation Limit		

Page 2 of 3

280 Newport Rd, Leola, PA 17540
Main 717.656.9326 ° Fax 717.656.0910
www.waypointanalytical.com

Project Information : Austin Unruh

Original Report Date : 12/30/2020
Revised Report Date: 12/30/2020
Received : 12/04/2020

Report Number : **20-339-0231**

REPORT OF ANALYSIS

Lab No : **91231**
Sample ID : **Wild**

Matrix: **Solids**
Sampled:

Test	Results	Units	MQL	DF	Date / Time Analyzed	By	Analytical Method
Moisture	**19.7**	%		1			AOAC 930.15
Dry Matter	**80.3**	%	0.10	1	12/07/20 15:19	KNP	AOAC 930.15
Crude Protein	**12.2**	% - dry	0.12	1	12/08/20 13:35	KNP	AOAC 990.03
Crude Fat	**2.09**	% - dry		1	12/21/20 13:00	RLF	AOAC 920.39
ADF Fiber	**25.0**	% - dry	0.124	1	12/08/20 10:22	RLF	AOAC 973.18
Neutral Detergent Fiber	**53.9**	% - dry	0.124	1	12/07/20 11:47	RLF	NDF
TDN	**69.4**	% - dry	0.124	1	12/08/20 10:22	RLF	AOAC 973.18
Estimated Net Energy (ENE)	**68.9**	ton/cwt - dry		1	12/08/20 10:22	RLF	AOAC 973.18
Net Energy Lactation (NEL)	**0.717**	Mcal/lb - dry		1	12/08/20 10:22	RLF	AOAC 973.18
Net Energy Maintenance (NEM)	**0.732**	Mcal/lb - dry		1	12/08/20 10:22		AOAC 973.18
Net Energy Gain (NEG)	**0.458**	Mcal/lb - dry		1	12/08/20 10:22		AOAC 973.18
Calcium	**0.56**	% - dry		1	12/08/20 11:30	KNP	FEED METALS
Phosphorus	**0.18**	% - dry		1	12/08/20 11:30	KNP	FEED METALS
Potassium	**1.49**	% - dry		1	12/08/20 11:30	KNP	FEED METALS
Magnesium	**0.11**	% - dry		1	12/08/20 11:30	KNP	FEED METALS
Sugar	**17.3**	% - dry		1	12/30/20 12:35	GLZ	AOAC 950.01

Qualifiers/ Definitions

DF	Dilution Factor	L	Limit Exceeded
MQL	Method Quantitation Limit		

Page 3 of 3

APPENDIX B

Recommended Reading

Gabriel, Steve. ***Silvopasture: A Guide to Managing Grazing Animals, Forage Crops, and Trees in a Temperate Farm Ecosystem.*** **Chelsea Green, 2018.** This book gives a wonderful overview of silvopasture. I especially like it for the insight on how silvopasture has been practiced around the world, and for the resources on how to thoughtfully thin a woods to create silvopasture. If you would like to manage your woods for silvopasture, I highly recommend this book.

Gerrish, Jim. ***Kick the Hay Habit: A Practical Guide to Year-Around Grazing.*** **Green Park, 2010.** This book has been a fundamental resource for me as I sought to understand how trees might serve graziers. Jim lays out the reasons to avoid making or feeding hay and lays out an alternative path forward. A must-read for any grass farmer who wants to make money.

Hay, Elspeth. ***Feed Us with Trees: Nuts and the Future of Food.*** **New Society, 2025.** A wonderful exploration of the history of nuts as staple crops, why we lost the culture around them, and why the time is ripe to bring them back.

Reid, Rowan. ***Heartwood: The Art and Science of Growing Trees for Conservation and Profit.*** **Melbourne Books, 2017**. Reid, an Australian "forester among farmers," lays out a beautiful vision for the use of trees on active farms, particularly in management for high-value timber. The book is beautiful enough for a coffee table and is packed with wisdom and insight from decades of planting, managing, and using trees. Though most of the trees highlighted are Australian, I highly recommend this book for anyone looking to manage their silvopasture with timber in mind.

Salatin, Joel. ***You Can Farm: The Entrepreneur's Guide to Start and Succeed in a Farm Enterprise.*** **Polyface, 1998.** A tremendous resource for any farmer, but especially for new and aspiring farmers. If you've already read it, do so again while keeping silvopasture and the untold number of new opportunities trees can unlock for your farm business in mind.

Savanna Institute. ***Perennial Pathways: Planting Tree Crops: Designing and Installing Farm-Scale Edible Agroforestry.*** **Savanna Institute, 2018.** This clear, well-written, and methodical guide will be very useful to anyone looking to add new tree crop enterprises like chestnuts, hazelnuts, or pecans to the farm. Available for free as a PDF through their website, or to purchase as a physical book.

Schwartz, Judith A. ***Cows Save the Planet: And Other Improbable Ways of Restoring Soil to Heal the Earth.*** **Chelsea Green, 2013.**

———. ***The Reindeer Chronicles: And Other Inspiring Stories of Working with Nature to Heal the Earth.*** **Chelsea Green, 2020.**

———. ***Water in Plain Sight: Hope for a Thirsty World.*** **St. Martin's Press, 2016.** A masterful storyteller, Schwartz relays the stories of farmers and land stewards restoring soils, ecosystems, and communities. The books are nothing short of inspirational and will give you motivation for the journey.

Shepard, Mark. ***Restoration Agriculture: A Real-World Permaculture for Farmers.*** **Acres U.S.A., 2013.** This is the book that first inspired me to start pursuing landscape-scale agricultural restoration using trees. Visionary and paradigm-shifting, this book is almost guaranteed to give you the itch to remake your landscape with trees.

Sibley, David Allen. ***The Sibley Guide to Trees.*** **Alfred A. Knopf, 2009.** This comprehensive and beautifully illustrated guide covers common and less-common trees throughout the country. For anyone looking to manage the trees already found on their land, a good tree identification guide is invaluable. You won't go wrong with The Sibley Guide to Trees.

Silver, Akiva. ***Trees of Power: Ten Essential Arboreal Allies.*** **Chelsea Green, 2019.** Akiva's simple, down-to-earth, and reverent writing makes this book both accessible and inspirational for folks of all backgrounds. I absolutely love the insightful profiles of ten "arboreal allies," including apple, poplar, mulberry, and black locust. I highly recommend this book to anyone interested in trees, especially for those who want to start propagating trees on their own.

Smith, J. Russell. ***Tree Crops: A Permanent Agriculture.*** **Harcourt, Brace, 1929.** Tree Crops has inspired countless people to create a permanent, two-story tree agriculture. It's as relevant now as it was when it was written way back in 1929, and wonderfully visionary. Shares how tree crops were being used around the world, including the most important ones to silvopasture like honey locust, persimmon, mulberry, and oak.

Stewart, Mart A. "From King Cane to King Cotton: Razing Cane in the Old South." ***Environmental History*** **12, no. 1 (January 2007): 59-79.** This article examines in depth the historical use of river cane throughout the South, including how fundamental it was to the raising of beef cattle

APPENDIX C

Worksheet

Step 1: Goals

1. What is your vision for the tree component of your farm? How would you like trees to transform your farm?
2. What are your main goals with silvopasture? Rank them, if you can, to highlight the most important goals to pursue.
3. When do you need additional feed the most?
4. Do you need more shade options? If so, what are the most pressing places on your farm?
5. Do you need more options for shelter and windbreak? If so, where and when?
6. wWould you like the trees to offer you new income or recreation opportunities? If so, list those.

Step 2: Species

1. What tree species are most likely to help you achieve your goals?
2. Which species are best to start with? Maybe they are inexpensive and easy, or simply readily available and the genetics don't matter as much.
3. What species, if any, are best to save until later, due to cost or lack of access?

Step 3: Establishment

1. How will you protect trees? Remember that they will need to be protected not only from livestock, but also from wildlife, including deer and rodents.
2. How will you reduce the competition with vegetation in the first few years?
3. How will you lay out your trees? Along fences? In grids? How far do you want them spaced?
4. Which areas should get first priority?
5. Sketch out or describe the first one or two phases of plantings.
6. What are the soil conditions in the areas to be planted?

7. How many trees will you need of each species for the next phase of planting? Use the tree calculation tables to help determine how many trees you need, then how many of each species you need.
8. Who will be in charge of planting Phase One? How many people will need to help? For reference, our crew averages planting a bit over three trees per person per hour when you factor in flagging, digging, planting, shelter installation, electric fencing and mulch.
9. What season will my next planting be in?
10. Who will be responsible for maintaining the trees?
11. Get dates on the calendar for the annual check-ins. We suggest around May and September.

Step 4: Funding

1. What funding sources will you pursue?
2. When will that funding likely be available?
3. What is your material (and labor, if needed) budget for your first planting phase?

Trees/acre calculation

	Sample Pasture				
Square feet per acre	43,560	43,560	43,560	43,560	43,560
Space between rows (feet)					
Space between trees in row (feet)	20				
Trees per acre	36.3				
Acres in pasture	5				
Trees needed	182				

Linear Feet Calculation

	Sample row						
Linear Feet in Row	1,200						
Space between trees in row (feet)	10						
Number of trees in row	120						

Species Needed

Honey Locust	
Persimmon	
Apple	
Oak	
Hybrid Poplar	
Hybrid Willow	
Black Locust	
Mulberry	
Total	

Budget

Item	Rate Per Item	Quantity	Total
Clonal trees			
Seedling trees			
Live stakes			
Tree shelters			
Mulch			
Labor per tree			
Total			

Supplies checklist for planting

- ☐ Trees
- ☐ Shelters
- ☐ Flags (optional)
- ☐ Vole guards
- ☐ Stake pounder (included when you buy shelters)
- ☐ Shovels or auger
- ☐ Mulch and means of moving it
- ☐ Measuring tape (to determine distance between trees and rows)

Checklist for Planting Day

- ☐ Each tree position staked out or flagged
- ☐ Stakes pounded in
- ☐ Holes dug
- ☐ Trees planted

- ☐ Vole guards installed (optional)
- ☐ Tubes installed
- ☐ Mulch installed
- ☐ Tree tubes protected as needed

Maintenance Steps

1. Is the tree alive, or does it need to be replaced?
2. Are there signs of damage/stress? If so, determine how to solve them.
3. Remove weeds inside tube.
4. After two to three years, remove stake if the tree is sturdy enough to stand on its own.
5. Remove tube once it is too small for the tree or is no longer needed to protect from rubbing.

Other Titles from Acres U.S.A. Books:

Dr. Paul Dettloff's Complete Guide to Raising Animals Organically

By Paul Dettloff

A hands-on reference for anyone at any level of experience who is interested in helping ruminants achieve optimal health naturally. Dr. Dettloff hands over a literal lifetime of learning and observation on how to raise livestock in healthier, more productive, and more humane ways. Comprehensive chapters present specific treatments for problems commonly seen with the digestive system, in reproduction with the respiratory, urinary and nervous systems, with skin, and in the circulatory and lymphatic systems. Handy go-to treatment summaries jump right to the how-to for hundreds of maladies for busy farmers, while the supporting text explains the underlying problem.

Restoration Agriculture

By Mark Shepard

Around the globe, most people get their calories from "annual" agriculture—plants that grow fast for one season, produce lots of seeds, then die. Every single human society that has relied on annual crops for staple foods has collapsed. *Restoration Agriculture* explains how we can have all of the benefits of natural, perennial ecosystems and create agricultural systems that imitate nature in form and function while still providing for our food, building, fuel and many other needs—in your own backyard, farm, or ranch. Based on real-world practices, this book presents an alternative to the agriculture system of eradication and offers exciting hope for our future.

Albrecht's Soil Fertility & Animal Health, Vol. II

By William A. Albrecht

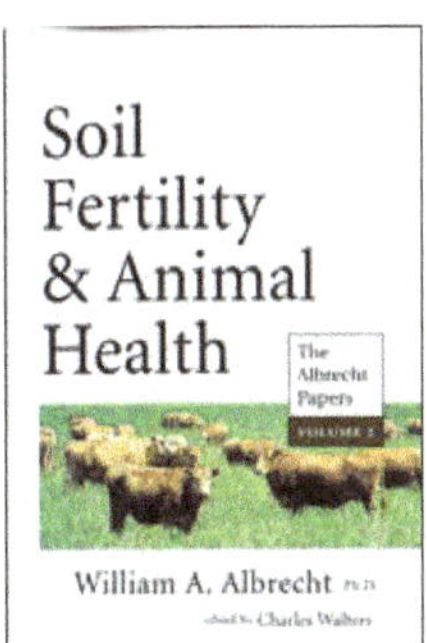

Albrecht was the premier soil scientist and was dismayed by the rapid chemicalization of farming that followed WWII. This book is a well-organized explanation of the relationship between soil fertility and animal and human health. This is a great book for those just familiarizing themselves with these concepts.

bookstore.acresusa.com

Visit acresusa.com
to learn about our magazine, events, books, and other educational opportunities.

www.ingramcontent.com/pod-product-compliance
Lightning Source LLC
LaVergne TN
LVHW061245100826
845148LV00008B/1026
* 9 7 8 1 6 0 1 7 3 9 8 9 6 *